# So Far As Is
# Reasonably Practicable

# So Far As Is Reasonably Practicable

## A guide to its meaning in occupational health and safety

David Farmer

BA LLB FInstPet MIOSH

Croner Publications Ltd
Croner House
London Road
Kingston upon Thames
Surrey KT2 6SR
Tel: 01-547 3333

This edition first published 1989
Reprinted 1989

Published by
Croner Publications Ltd,
Croner House,
London Road,
Kingston upon Thames,
Surrey KT2 6SR
Telephone 01-547 3333



While every care has been taken
in the writing and editing of this book,
readers should be aware that only Acts of Parliament
and Statutory Instruments have the force of law,
and that only the courts can authoritatively
interpret the law.

British Library Cataloguing in Publication Data

Farmer, David, *1928* –
So far as is reasonably practicable –
New ed. – (Croner health and safety guides)
1. Great Britain. Industrial health and
industrial safety. Law
I. Title
344.104'465

ISBN 1-85452-010-5

Printed by Whitstable Litho, Whitstable,
Kent

# Contents

# Introduction

Whenever an activity is the subject of any legal requirements those requirements, if they are to be understood and implemented, need to be as clearly stated as the language allows. With health and safety in the workplace it is possible to identify two clear categories of requirement.

The first is a positive, specific and exact one. A good example of this is to be found in s. 28 of the Factories Act 1961 where it is stated that "For every staircase in a building . . . a substantial handrail shall be provided and maintained". There is no room for doubt and no ambiguity about the requirement. For those who might perhaps wonder about the extent of the meaning of the word maintained, it is assigned a meaning in s. 176 of the Factories Act 1961. This says—"Maintained means maintained in an efficient state, in efficient working order, and in good repair". There is left no room for doubt.

The second category is a qualified and indeterminate one where the person upon whom the duty falls has to decide what needs to be done within the bounds of "reasonable practicability".

When the Health and Safety at Work, etc Act 1974 was passed its first General Duties section (s. 2) embraced all work and required that the health and safety of employees should be ensured "so far as is reasonably practicable". This connotes a stopping short of the absolute, using reasonable practicability as the yardstick against which to measure exactly how far one needs to go to achieve compliance with the letter of the law.

As will be seen later, reasonable practicability is a question of weighing-up the risks on the one hand and of balancing these against the resources called for to control them on the other.

This may appear to be a simple matter of exercising judgement, but in the absence of practical guidance in Approved Codes of Practice envisaged by s. 16 of the Health and Safety at Work, etc Act 1974, or from authoritative sources, it is by no means a simple matter at all. This book explores the whole question of how those with the statutory duty to ensure the health and safety of their employees and others, can go about it with the confidence that they have not overlooked anything vital.

Chapters 1–5 cover the background to the legislation, its contents and how it has been interpreted by the courts; and what guidance can be culled from official sources.

Chapters 6–9 cover a management strategy to review hazards by identifying exactly what they are; rating them for frequency and severity and then monitoring the standards and procedure introduced to control them.

# Chapter 1

## Background to present legislation

Prior to the Health and Safety at Work, etc Act 1974 a Committee on Safety and Health at Work (the Robens Committee) was asked to examine critically the content, administration and enforcement of occupational health and safety legislation in Great Britain and to make recommendations. The result was a Bill which, after some opposition to the use of the qualifying phrase "so far as is reasonably practicable" in debate, was passed into law with the phrase included.

Laws about conditions at work have been on our statute books for nearly two hundred years. The Health and Safety at Work, etc Act 1974 is a watershed. Before it the law was patchy in its scope and in its content and administration. After it employers have had to follow certain basic principles which govern all their activities at work, not only as they affect their employees but also with regard to anyone else who could be affected by those work activities. Before examining the present legal position a brief outline of the pre-1974 era is appropriate to set the scene.

## Early workplace legislation

The earliest legislation attempting to regulate working conditions applied to very few premises and was more an extension of the old Poor Law than a conscious attempt to regulate the actual workplace environment itself. The first Act was called The Health and Morals of Apprentices Act 1802 and it was introduced by Sir Robert Peel. It faced little Parliamentary opposition as a Bill and later, as an Act, it had very little impact in the country because it totally lacked any effective enforcement provisions.

This changed in 1833 when another Act of Parliament provided for the appointment of the first four factory inspectors and gave them substantial powers. At that time the only workplaces to which the law applied were textile factories and there were just over three thousand of them in the country.

Throughout the mid to late nineteenth century there were several separate statutes, each of which related to just one specific danger or hazardous process.

By the beginning of the twentieth century there were one hundred and fifty factory inspectors and just under a quarter of a million premises subject to their inspection. By this time a very much wider range of premises came within the scope of the law. It was now increasingly difficult for Parliament itself to find enough time to devote to the passage of Bills. To overcome this problem Ministers of the Crown were themselves specially empowered by legislation to make detailed regulations to cover the subjects dealt with by their Departments. The power to make "subordinate" or "delegated" legislation was extensively used.

Whatever else may be said about delegated legislation and its drawbacks it certainly gave rise to comments by the Committee on Safety and Health at Work – the Robens Committee Report (Cmnd. 5034) – set up in May 1970.

## Robens Committee Report

Lord Robens and six others, including representatives from industry, trade unions and academics, were appointed by the then Secretary of State for Employment and Productivity, Mrs Barbara Castle, to review the provisions made for the safety and health of persons at work. The exact terms of reference of the committee were:

> To review the provision made for the safety and health of persons in the course of their employment and to consider whether any changes are needed in:
> (1) the scope or nature of the major relevant enactments, or
> (2) the nature and extent of voluntary action concerned with these matters, and
>
> to consider whether any further steps are required to safeguard members of the public from hazards, other than general environmental pollution, arising in connection with activities in industrial and commercial premises and construction sites, and to make recommendations.

The Robens Report, as it is now customary to describe it, was presented to the Secretary of State for Employment, Maurice Macmillan, on 9.6.72.

The recommendations made were accepted by Parliament and are very largely responsible for the shape of our present legislation and the structure of the bodies determining its policy and responsible for its enforcement.

It is what the Robens report said about the background to the subject of this book that is now examined. Historically, said the report, we in this country had no abstract theory of social justice or the rights of man. We seemed to have been incapable of taking a general view. We found a particular evil in our workplaces and legislated for it and for it alone. "This practical, empirical approach has been associated with the development of high standards of safety and health protection, and with the attainment of a degree of systematic official supervision which is probably unsurpassed anywhere in the world." But, went on Robens, this source of strength was also a source of weakness because it resulted in a welter of detailed law constantly being extended and elaborated. Even so, the empirical approach could not keep pace with the continued rapid changes in processes and technologies. This resulted in us being left with much obsolescent detailed law and much new in industry not covered by the legislation at all. Basically there was too much law, concluded the committee, and people were heavily conditioned to expect external agencies to impose the detailed rules. "This attitude will not be cured so long as people are encouraged to think that safety and health at work can be ensured by an ever-expanding body of legal regulations enforced by an ever-increasing army of inspectors." Too little emphasis is put on "personal responsibility and voluntary, self-generating effort."

## Recommendations of the Robens report

The Robens report said that an effective self-regulating system was needed and that an enabling Act should be passed which should begin by enunciating the basic and overriding responsibilities of employers and employees. "This central statement", it continued, "should spell out the basic duty of an employer to provide a safe working system including safe premises, a safe working environment, safe equipment, trained and competent personnel, and adequate instruction and supervision".

The Committee had heard evidence that such a statement would be "too general to be meaningful and helpful in practice". It did not accept this. Such a statement, would make it clear to people that an "all-embracing responsibility, covering all workpeople and working circumstances" existed and would apply whether or not a particular matter was covered by a specific regulation. It is clear that despite certain legal experts arguing that the common law already provided such a declaration of principles, the members of the Committee saw much to commend the statement of the broad principles actually being in the Statute itself. Their recommendation was accepted. We now have it in s.2 of the Health and Safety at Work, etc Act 1974, which, as we know, contains the phrase "so far as is reasonably practicable". It is repeated in the section no less than six times.

**New regulations**

> Regulations which lay down precise methods of compliance have an intrinsic rigidity, and their details may be quickly overtaken by new technological developments. On the other hand, lack of precision creates uncertainty. This is a problem to which our attention was repeatedly drawn during the course of the inquiry. The need is to reconcile flexibility with precision. We believe that, wherever practicable, regulations should be confined to statements of broad requirements in terms of objectives to be achieved.

The Committee commended this approach by referring to the Alkali Works Regulation Act 1906 and the use of the phrase best practicable means referred to later (page 20).

## Health and Safety at Work Bill

In May 1974 a Standing Committee of the House of Commons let it be known through the pages of its report what MPs thought about the phrase "so far as is reasonably practicable". Mr Patrick Mayhew said it was a formula well understood in the courts.

Mr Bob Cryer said the words were of great importance, being regarded by many as a lawyer's charter. He believed the phrase reduced the level of safety legislation. Amendments were accordingly put down to leave out the word "reasonably" where it preceded the word "practicable" in eleven places where it was written into the Bill. This would have had the effect of making the employer's duty very much more stringent by removing all elements of flexibility.

"If we impose an absolute duty upon the employer to provide a safeguard against any risk, everyone bends his energies to see that everything which can conceivably be done is done in order to protect a man from that risk." Otherwise, continued the report, citing an earlier debate, "When we say to an employer 'Do all that is reasonably practicable' he carries on as he has always done. If there is an accident he argues 'The precautions I took were reasonably practicable. I have done what all other employers do in the like situation'". That sort of argument, as the common law cases show, carries great weight in the courts, said Bob Cryer. He went on to declare "The words 'reasonably practicable', in my view, diminish the rights, powers, and opportunities of the Factory Inspectorate". Only if the law is precise is prosecution undertaken. If it is imprecise the inspectors try to use persuasion and it fails, continued Mr Cryer. "When we legislate we should be specific and lay down firm protections. We ought not to say that there is a price on life."

When in the debate someone asked why the Mines and Quarries Act 1954 had had "reasonably" deleted from it when it was a Bill going through Parliament, it was explained that the Mines and Quarries legislation was specific but the HSW Bill was not. The former could contain absolute requirements because of its specific nature, but the latter could not because of its universal, general, and all-embracing nature.

Mr Weitzman's contribution to the debate was to reinforce the view that omitting the word "reasonably" would in many cases "make work not physically possible". He explained why he believed the anxieties expressed were "ill-founded and based on a misconception of what the position really is". He explained that there were three standards. The first is where a duty is absolute and "any failure is ipso facto proof of a breach". The second is the standard calling for a precaution "unless it is impracticable", or so far as "practicable", where practicable means "physically possible in the light of current knowledge and invention". The third is the standard of reasonable practicability. "In general" said Mr Weitzman, "it is not appropriate to apply the first and second standards of obligation in non-specific situations."

The Solicitor-General made three comments on the issue:

(a) To remove the word "reasonably" would deny someone accused of a criminal offence the opportunity of arguing that what he or she did was reasonable and hence should not be branded as criminal.
(b) To make a requirement too stringent may give rise to a reluctance to enforce the provisions by the authorities

concerned. Especially when the numbers of inspectors are less and their workload has increased.

(c) To make a general provision subject to reasonable practicability does not mean Ministers should not direct their minds to specific dangers and make specific regulations to deal with them: *they should very definitely follow such a course.*

The Minister attending the Committee, Mr Harold Walker, said he could not accept the proposed deletion of the word "reasonable" and the amendments suggesting that they should be withdrawn wherever they appear.

That part of our law making process which debates a Bill clause by clause in committee, looked at the proposed amendments, probed their meaning and their practical effects, and decided to use the phrase "so far as is reasonably practicable" as an integral part of the general statement of an employer's duty which the Robens Committee on Safety and Health at Work had recommended.

The courts had much earlier, in 1949, looked at the same phrase in an actual case before them and this is the aspect of the subject to which we now turn. We shall find that it creates as many problems as it solves.

# Chapter 2
## The Health and Safety at Work, etc Act 1974

The Health and Safety at Work, etc Act 1974 has been far reaching in its effects, not least because it states for the first time a general principle by which every employer, employee and the self employed are bound. Most of the duties the Act imposes however are not absolute. They are qualified by the words "so far as is reasonably practicable". It is necessary to know where the qualification appears and who and what is affected by it.

## "It shall be the duty"

"It shall be the duty of every employer to ensure, so far as is reasonably practicable, the health, safety and welfare at work of all his employees."

Many readers will be familiar with this sentence from s. 2(1) of the Health and Safety at Work, etc Act 1974 but most will inevitably be uncertain about its exact meaning. They will not be sure of the effects on their statutory obligations of the six words "so far as is reasonably practicable". This is worrying, simply because what is at issue is a legal duty, failure to comply with which may lead to criminal proceedings and the imposition of penalties by the courts, or of the service of improvement or prohibition notices by enforcing authority inspectors.

This book has gathered together information about the concept of reasonable practicability. It reviews what the courts and others have said about it. It also suggests courses of action which anyone can follow in order to satisfy themselves and others, when the need arises, that they are doing what the law requires of them. Clearly, before looking at anything outside the Health and Safety at Work, etc Act 1974 which has been written or said about the phrase so far as is reasonably practicable, it is essential to look further at the statute itself.

## S.1

Looking at the first section of the Act it is seen that its purpose is to enable old statutes and regulations to "be progressively replaced by a system of regulations and approved codes of practice operating in combination with the other provisions of this Part (ie Part 1 of the Act) and *designed to maintain or improve* the standards of health, safety and welfare established by or under those enactments" (author's italics). Pointing this out is important because many critics of the qualification so far as is reasonably practicable regard it as a formula for diminishing or diluting the effectiveness and stringency of occupational health and safety law. The extract above gives the official lie to such accusations. It would also make any regulations or approved codes of practice that did not maintain or improve standards ultra vires and open to challenge in the courts. Ultra vires is the expression used about anything done in excess of an authority conferred by law and therefore an invalid act.

## S.2

S.2, which states the general duties of employers to their employees (and sub-section 1 of which was quoted in full at the beginning of the chapter) uses the phrase so far as is reasonably practicable six times in relation to:

(a) the provision and maintenance of plant and systems of work;
(b) arrangements for ensuring safety and absence of risks to health in connection with the use, handling, storage and transport of articles and substances;
(c) the provision of such information, instruction, training and supervision as is necessary to ensure the health and safety at work of employees;
(d) the maintenance of any place of work under the employer's control in a condition that is safe and without risks to health and the provision and maintenance of means of access to and egress from it that are safe and without such risks;
(e) the provision and maintenance of a working environment for employees that is safe, without risks to health and adequate as regards facilities and arrangements for their welfare at work.

## Ss.3 and 4

S.3 of the Health and Safety at Work, etc Act 1974 sets out the general duties of employers and the self-employed to persons other than their employees. It again qualifies these duties by the phrase so far as is reasonably practicable. S.4 of the Act applies broadly similar duties to s.3 in relation to those who are not employees, but who "use non-domestic premises made available to them as a place of work, or as a place where they may use plant or substances provided for their use there". The duties relate to access and egress, plant and substances. Again the extent of the duty is qualified by the reasonable practicability clause.

In the case of *R v Mara* [1987] 1 WLR 87 it was held that an undertaking is still being conducted whether the company's own employees are there or not. As Mr Mara discovered, it is not possible to treat s.3 as applying only when the undertaking is *actually in the process of being carried on.* The case related to the electric cable of a floor scrubbing machine which was damaged and unsatisfactorily repaired. As a consequence of this shortcoming an employee of the stores who, with Mr Mara's knowledge, was using the machine, was electrocuted. Mr Mara admitted that he knew the cable was taped over to cover its damaged parts and also that copious quantities of water were used in cleaning the stores loading bay area. Mr Mara's initial contract was with the stores for cleaning both shop and loading bay. The constant arrival of delivery vehicles made cleaning of the loading bay by Mr Mara's employees difficult and so he reached a new arrangement to make the floor scrubbing machine available to employees of the stores so that they could clean the loading bay themselves when it was convenient for them to do so. Mr Mara contended that on the day of the accident, a Saturday, his employees were not present and hence he was not conducting his undertaking and therefore could not be subject to the requirements of s.3. The court did not agree with this contention.

## S.5

S.5 of the 1974 Act does not use the phrase so far as is reasonably practicable to qualify the extent of the duty but uses another criterion against which the performances of those having duties is judged. The duty is to use the "best practicable means" and it relates to the prevention of the emission into the atmosphere of noxious or offensive substances and for rendering such emissions harmless and inoffensive. The philosophy behind best practicable means has been in use for over a century. A chief alkali inspector likened it to an "ever-tightening elastic band" by which emission levels could be reduced as knowledge and technological progress improved. The same man attached great importance to consultation with industry before deciding on the best practicable means and once the formula was agreed it was used to enforce compliance. Such a philosophy of consultation is part and parcel of current occupational health and safety practice. It is mirrored in the legislation and in the formal structure of the Health and Safety Commission (HSC).

The consultative process also involves various Industry Advisory Committees appointed by the Health and Safety Commission which have been likened to "mini" health and safety commissions for their industries. It is certainly highly probable that the courts would pay very close attention to the nature of the guidance and standards incorporated into the many publications for which the Industry Advisory Committees have been responsible. Employers are likely to be readily deemed to know what the Committee for their own industry has suggested as the appropriate method of dealing with their industry's problems. In deciding what is likely to be held to be reasonably practicable Industry Advisory Committee publications are sure to play a particularly significant role.

## S.6

S.6 puts duties on designers, manufacturers, importers and suppliers in relation to articles and substances used at work. Many of the duties imposed in the areas of design, construction, research into hazards and erection and installation of plant are qualified ones. Again the qualifying phrase used is so far as is reasonably practicable.

One aspect of the duties contained in s.6 related to the rather passively expressed duty, placed on the designer, manufacturer, importer or supplier, to make adequate information available to the user. The information should relate to the designed use of the article, tests to which it has been subject, and any conditions to be observed to enable it to be used safely and without risks to health.

Similarly with substances, the user should have the information necessary to ensure that the substances can be used safely. Many manufacturers have made available data sheets containing all the necessary information for users; others have not.

When s.36 and schedule 3 of the Consumer Protection Act 1987 came into force on 1.3.88, their effect was to bring into operation a number of amendments to s.6 of the Health and Safety at Work, etc Act 1974. The passive duty to make information available was revoked and was changed to a positive duty to secure that persons supplied with articles and substances *are provided with adequate information about matters pertinent to health and safety.* Furthermore, so far as is reasonably practicable, the persons so supplied with adequate information must also be provided with "all such revisions of information... as are necessary by reason of its becoming known that anything gives rise to a serious risk to health and safety". It is submitted that the use of the phrase so far as is reasonably practicable in the s.6 amendment does not lend itself to the interpretation discussed later in Chapter 3, but that the words are used in their literal sense to mean that the supplier must expend reasonable effort to acquaint customers with new information relevant to the issues of health and safety, as and when it comes into his or her possession. This change in the law will almost certainly be most significant in the area of keeping customers up to date with new data about the health effects of the substances supplied.

By now it will have become abundantly clear that not only is the qualification of "reasonable practicability" applied generally in s. 2(1) but more specifically elsewhere. There is, however, no mention of what the phrase means in practice. Those who seek enlightenment in s. 53 of the Health and Safety at Work, etc Act 1974 which gives interpretations and definitions will be disappointed, because there is none for this phrase. We have to look at matters of judicial interpretation which is covered in Chapter 3.

## Approved Codes of Practice

Some relief and encouragement is at last to be found in s.16 of the 1974 Act. For this is the section which gives the Health and Safety Commission the power to issue and approve codes of practice. The purpose of a code of practice is to provide "practical guidance with respect to the requirements of any provision of ss. 2 – 7, or of health and safety regulations, or of any of the existing statutory provisions". In other words, this phrase covers any current or future statutory requirements in occupational health and safety. The legal status and effects of Approved Codes of Practice (at the time of writing approaching 40 codes of practice have been approved) is dealt with later. At present it is sufficient to say that a code of practice includes "a standard, a specification, and any other documentary form of practical guidance" (s.53 Health and Safety at Work, etc Act 1974). At last, some flesh to put on the bones of the general duty.

## Personal responsibilities

It may have been observed that in our review of the early sections of the 1974 Act no reference was made to s.7 which prescribes the duties of an employee. Yet s.7 is included among those for which codes of practice may be approved by the Health and Safety Commission. It is interesting to observe that an employee's duty is not qualified by the phrase so far as is reasonably practicable. An employee's duty is to take "reasonable care for the health and safety of himself and of other persons who may be affected by his acts or omissions at work". The employee also has a duty to co-operate with others as far as is necessary to enable duties or requirements to be complied with.

In the absence of the reasonable practicability qualification, for any offence alleged against an individual under ss.7 or 8 (s.8 relates to the duty on all persons not to intentionally or recklessly interfere with or misuse anything provided in the interests of health, safety or welfare) it is for the prosecution to prove in the conventional way that what it alleges against the person charged is so "beyond reasonable doubt".

If, however, an employee is a director, secretary or other similar officer of the company (or "body corporate" as s.37 of the Health and Safety at Work, etc Act puts it) he or she may also be charged with an offence, as well as the company, if it is proved that the offence was committed with his or her consent, connivance, or to have been attributed to any neglect on his or her part. So reasonable practicability may be an issue for an individual who holds any of the above offices or, for that matter, anyone who purports to act in the capacity of any such officer.

In the interpretation of the phrase "or other similar officer" it has been submitted that to be on a par with the terms director, manager or secretary only the most senior officials in a company hierarchy would be covered. The issue was pursued in the case of *J Armour v J Skeen (Procurator Fiscal, Glasgow)* [1977] IRLR 310. A director of roads was prosecuted for not writing a safety policy.

## Onus of proof reversal

One final section of the 1974 Act needs to be examined closely. It is s.40. This section states that in any proceedings for an offence consisting of a failure to comply with a duty or requirement to do something so far as is reasonably practicable it is for the accused to prove that it was not reasonably practicable to do more that was in fact done to satisfy the duty or requirement.

It has been suggested that the effect of s.40 is not the soft option for inspectors which it may appear to be. Inspectors are, by virtue of s.38 of the Health and Safety at Work, etc Act 1974, the only persons empowered to take proceedings under the Act except by or with the consent of the Director of Public Prosecutions. The courts, it has been submitted, will need to be convinced by the inspector that there is a case against the person or persons charged and that only at this juncture will it fall to those accused to defend themselves, if they so wish, by arguing that they did do all that was reasonably practicable.

Looking at this in one way it could be argued that the reversal of the traditional onus of proof here from the prosecutor to the accused is the last straw. On the other hand, it is no crime not to follow what a code of practice says as long as you are in the position to satisfy a court that the methods of work you chose were as good as, or better than, the one prescribed in the approved code.

Any provision of a code of practice is admissible in proceedings when it appears to the court to be relevant to the requirement or prohibition alleged to have been contravened. If it is proved at any material time that there had been a failure to observe any provision of an approved code "which appears to the court to be relevant to any matter which it is necessary for the prosecution to prove in order to establish a contravention of that requirement or prohibition, that matter shall be taken as proved unless the court is satisfied that the requirement or prohibition was (in respect of that matter) complied with otherwise than by the way of observance of that provision of the code".

What have we learned so far? We find that a statute creating duties, of the most far-reaching kind, in all places of work, is not exact, clear, or precise as it stands. We find that other subordinate legal and quasi-legal creatures (regulations and codes of practice) also use the same qualifying phrase. Hence their meaning may also be difficult to determine in the exact and practical terms necessary to those having to comply with duties. The law has to be translated into positive action. How can this be done if the law appears to be so vague and imprecise?

# Chapter 3

## Interpretation of legislation by the courts

The phrase "so far as is reasonably practicable" was not invented for use in the Health and Safety at Work, etc Act 1974. It had been used earlier in legislation and had been interpreted judicially well before the 1970s. Since the 1974 Act some further indications of the extent of its meaning have emerged.

## Interpretation hierarchy

Whenever courts adjudicate on a case brought before them, the judgments handed down contain two elements. The first is called the ratio decidendi, or the ground of decision. It is this which makes the decision a binding precedent for the future and becomes a principle of law. It can be overturned only by a higher court on appeal, or by Parliament which can legislate and change the law by passing a statute saying what the new law is to be. If the decision is one made by the House of Lords, there is no higher court to overturn its decision and then only Parliament may make any changes. Of course, only if a case is on all fours with the earlier case creating the precedent will the court follow and apply the decision in that earlier case. Much of the skill of advocacy lies in persuading judges that a case before them is not on all fours with an earlier case and that they should therefore not follow the precedent.

The second element of a judgement, which need not involve us further here, is the obiter dicta or "things said by the way". Such observations are not binding as precedents.

## The judgement in *Edwards v NCB*

The present judicial authority ruling on the meaning of so far as is reasonably practicable is to be found in *Edwards v National Coal Board* [1949] 1 KB704, [1949] 1 All ER 743, 65 TLR 430, CA.

The following appears in the report on the case:

> Reasonably practicable is a narrower term than physically possible, and implies that a computation must be made in which the quantum of risk is placed in one scale, and the sacrifice, whether in money, time or trouble, involved in the measures necessary to avert the risk, is placed in the other; and that, it if be shown that there is a gross disproportion between them, the risk being insignificant in relation to the sacrifice, the person upon whom the duty is laid discharges the burden of proving that compliance was not reasonably practicable. This computation falls to be made at a point of time anterior to the happening of the incident complained of.

## The facts

Mr Edwards was a colliery timberman employed down a mine in South Wales. He was killed there in an accident in November 1947. He died when a latent defect in the rock strata at the side of an underground pit roadway led to a fall of rock. A fossilised tree about which no-one was aware at the time Mr Edwards' accident gave rise to a "glassy slant" in the side of the roadway. The expression "glassy slant" is used to describe a stratum which is hard, shiny and slippery and because it does not bind with adjacent material it is likely to slide off without warning if pressure is put on it.

Evidence was given that the presence of this hazard could not have been discovered by superficial examination. Nor was its existence to be anticipated. Other evidence that earlier subsidence nearby and shotfiring contributed to the fall was not accepted by the court. On the evidence put to him, the judge thought it was not reasonably practicable for the employer to prop and line every inch of his underground roadways. The judge also agreed with the employer who said he would be faced with an impossible burden if his duty was to protect every part of every roadway in the absence of some indication that protection was necessary. In a world of unlimited resources such a standard would be the perfect answer. In reality it was not an economically feasible way of continuing to operate. If there had been a history of similar incidents over the years it would have been different but it appeared that this was not the case.

Although the judge said that cost was an essential element to be considered, he added a balancing perspective by saying that the greater the risk that could be shown, the less weight could be attached to consideration of the costs involved in offsetting the risks.

## Affirmation by the House of Lords

The *Edwards* case was referred to with approval in *Marshall v Gotham Co Ltd* [1954] AC 360 [1954] All ER 937. In this second case, the House of Lords were again considering an incident in a mine. It was a gypsum mine and again a collapse attributed to a rare geological fault called a "slickenside". In their judgement, which was considering the requirement to make a gypsum mine roof safe so far as is reasonably practicable, their Lordships emphasised the rarity of the phenomenon giving rise to the danger (not known in the mine for twenty years). Furthermore, the House of Lords said it was the known risk which had to be taken into account and balanced against safety precautions. Lord Reid said:

> if a precaution is practicable it must be taken, unless in the whole circumstances that would be unreasonable. And, as men's lives may be at stake, it should not lightly be held that to take a practicable precaution is unreasonable... the danger was a very rare one. The trouble and expense involved in the use of the precaution, while not prohibitive, would have been considerable. The precautions would not have afforded anything like complete protection against the danger, and their adoption would have had the disadvantage of giving a false sense of security.

It is probable that their Lordships would have found against the defendant, regardless of expense, if the value of systematic roof support had been established. They had heard that such support *would not have prevented the fall,* though it might have lessened the chances of it happening to someone's detriment.

Taking a closer look at the facts of these two key judgements we find that in both instances the precipitating causes were natural and not manmade in origin; that latent defects in the fabric of the mine, incapable of detection by superficial examination, were responsible (by superficial it is not meant that the examination was slapdash or careless but that only the surface was capable of being looked at); and that the precipitating causes were rare.

Despite this, the test of what constitutes reasonable practicability stands, but how it can be related to everyday management decisions "anterior to the happening" (when it has to be made according to the judgement) is the vital issue and is the subject dealt with later.

Meanwhile there are some other judgements of significance which need to be mentioned here, before passing to a consideration of the line taken by HSC/HSE in interpreting "reasonable practicability" as revealed in some of their publications.

## Practicability

Although we have seen that designers and manufacturers are under certain duties to undertake research and discover, eliminate or minimise risks to health and safety in articles and substances, users are limited in their duties by the present state of knowledge and invention.

In the case of *Adsett v K & L Steel Founders and Engineers Ltd* [1953] 1 All ER 97 the plaintiff sued the company, arguing that the employer had not taken all practicable measures to stop him inhaling dust. Mr Adsett had developed pneumoconiosis while doing his job. The job consisted of shovelling various grades of casting sand through a grating onto a conveyor below. Atmospheric contamination by silica dust resulted. As soon as the idea of local exhaust ventilation was established it was implemented by the defendants but too late to prevent Mr Adsett's disabling condition.

The relevant section of the Factories Act 1961 (s.63) was the relevant section until revoked by Regulation 19 and Schedule 8 of the Control of Substances Hazardous to Health Regulations 1988 (COSHH) called for all practicable measures to be taken to protect employees from the inhalation of dust.

The use and effect of the phrase "so far as is reasonably practicable" in the COSHH Regulations is covered at some length later.

On its own, the word practicable is stricter in its application than reasonably practicable and the question of cost is irrelevant. Nevertheless because no measure could be practicable if it was not within "current knowledge and invention" the court held the employers not in breach of their statutory duty. On appeal it was argued that the technology to apply appropriate extraction ventilation did exist but that neither the defendants (nor their peers in the industry) had thought of applying it in a way which would have helped the plaintiff. The Court of Appeal said that to be practicable a precaution had to be known for the particular application in question by the industry and by the relevant experts. Mr Adsett lost his case.

## Restrictions and extensions

Within the last decade there have been very few judgements

which materially affected the main authorities described above. Six are considered worthy of mention but only two establish binding precedents. Only two have any far-reaching practical significance. The other four cases are merely indicators of the responses of the tribunals and courts hearing the cases brought before them by enforcing authority inspectors. In no case does the decision reported do other than demonstrate a line of thinking.

### Particular solutions

It is not up to an enforcing authority's inspectors to seek to impose particular solutions to problems by saying what they want done in improvement notices which they serve. All the circumstances have to be considered.

Following an incident at a roller truck an inspector issued an improvement notice in which he ordered the company to provide free safety shoes because the roller truck had injured an employee's foot by running it over. The company, who already made protective footwear available to its staff at cost (which it recovered from them at £1 per week), appealed. The order in the notice would be too costly and disproportionate to the risk said the company, who claimed their practice was in line with normal trade practice and was in all the circumstances reasonable. One swallow doesn't make a summer, and one accident doesn't condemn a machine out of hand and call for disproportionate precautionary measures. Every case is judged on its merits, *Associated Dairies Ltd v Hartley* [1979] IRLR 171.

### Alternative solutions

An inspector served a notice on a building society to erect bandit screens to protect staff from the potential dangers posed to them by armed thieves entering the premises. On appeal to an industrial tribunal the notice was upheld. The building society appealed to the High Court who allowed the appeal. The duties under s.2 of the Health and Safety at Work, etc Act 1974 were not absolute, said the Queen's Bench Division, so the industrial tribunal in determining whether there had been a contravention of those duties had to consider whether it was reasonably practicable to provide screens at the offices referred to in the notice. To do this the risk had to be weighed against the precautionary measures necessary to avert the risk. The tribunal had erred in law and the notice was quashed. Other factors commented on were the views of the society that screens were contrary to their desired image of friendly informality and confidence building in would-be customers. Moreover staff were trained, if threatened with violence, to offer no resistance to thugs. There are more ways of saving a threatened cashier than by putting him or her behind armoured plateglass. *West Bromwich Building Society Ltd v Townsend* [1983] ICR 237, IRLR 147.

## Extent of duty

Part of the system of legal controls prior to the making of the Control of Substances Hazardous to Health Regulations 1988, ie prior to 1.10.89, was connected with the official publication of Occupational Exposure Limits (OELs). These had previously been referred to as Threshold Limit Values (TLVs) and consisted of control limits and recommended limits. The lists of data were published, and brought up to date annually in an HSE Guidance Note EH40 "Occupational Exposure Limits".

These limits are figures indicating concentrations of airborne contamination of the workplace by dust and fumes. "They are limits which have been judged after detailed consideration of the available scientific and medical evidence to be 'reasonably practicable' for the whole spectrum of work activities in Great Britain." It is apparent that the enforcing authorities' view is that even after achieving the degree of control of atmospheric contamination indicated by the appropriate Occupational Exposure Limit, further action may be necessary to fulfil the requirements of the Health and Safety at Work, etc Act 1974 and the Control of Substances Hazardous to Health Regulations 1988. In other words, doing what is reasonably practicable does not cease to be a statutory obligation just because a degree of control within an officially published figure can be shown. Why should this be so? Is it not unfair to require control below a stipulated exposure limit and when this is achieved to require yet greater effort to reduce the contamination still further? It is submitted that it would be manifestly unfair if the exposure limit represented a *safe limit*. It is not a safe limit, however. It is a compromise limit set within achievable bounds in the opinion of those considering the issues and the hazards and risks involved.

Two cases brought before a Stipendiary Magistrate by one of H M Inspectors of Factories indicate the above principle clearly. The facts are, very briefly, that during asbestos removal by a contractor at a dock car park, atmospheric contamination was above the relevant Control Limit for the type of asbestos concerned. Both the contractor and the dock company were prosecuted. Against the dock company it was said that even though it could not be proved their employees worked in dust levels above the Control Limit, a risk to the health of their employees working in the car park did exist. Moreover, the dock company could have provided barriers and instructions to prevent their employees entering the car park, or they could have provided their employees with half-mask respirators. Such provisions were "reasonably practicable". *Brearey v Anthony Kelly* [1984], *Brearey v Harwich Dock Company Limited* [1984] (Unreported).

## Looking at the knowledge of others

Almost no work activity is carried on in total isolation all of the time. Other people from outside the particular premises concerned become involved. They work in positions so close to employees at those premises that if they do dangerous things not only are they themselves at peril, but so too are the employees of the occupier of the premises visited.

Doing what is reasonably practicable includes providing information of hazards to employees of *other* employers working alongside an employer's own people. He or she must ensure that the ignorance of others' employees does not endanger his or her own.

The facts giving rise to the Court of Appeal ruling were as follows. During the fitting-out of a ship, *HMS Glasgow*, oxygen leaked into confined parts of that ship and the atmosphere became oxygen enriched. The leak came about because an employee of a sub-contractor failed to disconnect an oxygen hose as he should have done before ending work for the day. Next morning a welder's torch caused a fierce fire during which people were killed.

The shipbuilders submitted they were under no duty to subcontractors' employees, ie to tell about oxygen enrichment hazards. The trial judge ruled they were. On appeal the Court of Appeal affirmed the ruling in the lower court.

If the provision of a safe system of work for the benefit of an employer's own employees involves giving information and instruction on potential dangers (ie oxygen enrichment) to persons other than his or her own employees, then that employer is under a duty to give information and instruction to such other persons, so far as is reasonably practicable. *R v Swan Hunter Shipbuilders Limited* [1982] 1 All ER 264; [1981] ICR 831; [1981] IRLR 403 CA

## Breach self-evident

A lift used at a Tesco store in Stoke-on Trent was involved in an accident in which a part-time assistant was severely burned when he received an electric shock. His hand came into contact with live electric parts in the area behind the control panel. He was able to come into contact with these parts because the panel, which had provision for four securing screws, had at least two and probably three of them missing at the time of the incident. From the fact that the screws were missing it was self-evident that the state of the lift installation constituted a breach of s.2, in that the employers had not, so far as reasonably practicable, ensured the safety of employees. *Tesco Stores Limited v Seabridge* [1988] The Times, 29 April, QBD.

## Burden of proof

In Chapter 3 reference was made to s.40 of the Health and Safety at Work, etc Act 1974 which puts the onus of proof in questions of reasonable practicability upon the person against whom there is levelled an allegation of not doing all that is, or was, "reasonably practicable". The reversal of the norm is not quite as bad as it may appear because the *burden of proof* is substantially less. The courts have established that the burden of proof is discharged in these cases if the evidence produced by a person accused of not doing all that is reasonably practicable justifies the conclusion that the *balance of probabilities* is in his or her favour. This is a burden which is no greater than that resting upon a party to a civil action. It is enough if the accused can satisfy a court of the probability of that which he or she is called on to establish. It is equally enough if an employer can satisfy an enforcing authority inspector to the same degree. The inspector, as prosecutor in all other cases than the type described here, has to prove in any legal proceedings that what he or she alleges is *beyond reasonable doubt*. This is a much more stringent burden.

Of course, it has to be recognised that most enforcing authority inspectors display an "Oliver Twist syndrome": they will tend to ask for more. This is not a criticism of them but a recognition that they display a natural and laudable wish to see improvements in working conditions without themselves having to worry about the availability of resources to bring them about by the expenditure of hard-won funds obtained in competition with others not perhaps similarly constrained.

## Duty to ensure the safety of non-employees

Those who, to any extent, have control of premises have certain duties, under s.4 of the Health and Safety at Work, etc Act 1974, towards non-employees who work on those premises. In *Austin Rover Group Ltd v H M Inspector of Factories* [1989] 3 W L R 520; [1989] 2 All E R 1087 the House of Lords examined the nature and extent of this duty. The House of Lords ruled that the company had, so far as was reasonably practicable, taken such measures as it was reasonable for it to take in the circumstances of the case. The circumstances were that a contractor's employee was killed while he was undertaking cleaning work at a paint spray booth. The accident followed from the contractor failing to follow certain safety procedures agreed with Austin Rover Group Ltd. It was held that since the contractor (Westleyshire Industrial Services Ltd, which pleaded guilty to a charge under s.2 of the Health and Safety at Work, etc. Act 1974 and was fined £2000) failed to abide by its agreement, it could not be said that it was reasonable to expect Austin Rover Group Ltd to have taken measures to make the premises safe against such unanticipated misuse of them.

# Chapter 4
## Detailed analysis

Although being careful not to usurp the interpretative role of the courts, various official publications since 1974 have indicated how the Health and Safety Commission and the Health and Safety Executive themselves regard the ubiquitous phrase. An understanding of the official view is clearly an advantage to those whose duties are qualified by it.

## Published comments

We have now seen what the Health and Safety at Work, etc Act 1974 (HSWA) says, and how the phrase so far as is reasonably practicable appears in various places in it.

We have examined how the Robens Committee regarded the general need for a statement of broad principle to inspire and direct the action of employers in the safeguarding of their employees and others. We have seen too that to avoid setting too arduous and absolute a standard of duty for general purposes (despite some opposition in debate in the House of Commons), the Government of the day decided to use the qualifying term already known and used by legislators previously. In the last chapter we saw how the courts had evolved a test to establish what reasonable practicability means and how some other decisions had affected matters.

Now, in this chapter, we turn to the various places where the Health and Safety Executive have given some indication and guidance on how they regard the phrase and how those to whom it applies should behave in their workplaces. Although these references have no legal standing per se, it is prudent to be able to gauge the likely attitudes and stance of enforcing authorities from whom any initiatives leading to legal proceedings will stem.

There will be no probing of the meaning of so far as is reasonably practicable by the civil courts involving ss.2 – 7 of the 1974 Act, because s.47 specifically states that no right of action is conferred in any civil proceedings respecting any failure to comply with the sections cited.

### A Guide to the Health and Safety at Work Act – Health and Safety series booklet HS(R)6 (HMSO)

*Paragraph 22:* Although some of the duties imposed by the Act and related legislation are absolute, many are qualified by the words "so far as is practicable" or "so far as is reasonably practicable".

*Paragraph 23:* Although none of the expressions is defined in the Act, all three have acquired quite clear meanings through long established interpretations by the courts.

Someone who is required to do something so far as is reasonably practicable, must assess, on the one hand, the risks of a particular work activity or environment, and, on the other hand, the physical difficulties, time, trouble and expense which would be involved in taking steps to avoid the risks. If, for example, the risks to health and safety of a particular work process are very low, and the cost or technical difficulties of taking certain steps to avoid those risks are very high, it might *not* be reasonably practicable to take those steps. However, if the risks are very high, then less weight can be given to the cost of measures needed to avoid those risks. The comparison does not include the financial standing of the employer. A precaution which is "reasonably practicable" for a prosperous employer is equally "reasonably practicable" for the less well off. The phrase so far as is reasonably practicable, without the word reasonably, implies a stricter standard.

*Paragraph 24:* If someone is prosecuted for failing to comply with a duty so far as is reasonably practicable, it is up to the accused to show the court that it was *not* practicable or *not* reasonably practicable (as appropriate) for him to do more than he had in fact done to comply with the duty. Note: The paragraph above also commented in like vein on the interpretation of the phrase "best practicable means" (BPM) but these references have been omitted for clarity.

This commentary in the standard guidance published by HSE in 1980, does not help very much in practical terms. In any event whatever may be gleaned from the text is partly weakened by the traditional, general caveat saying "it is not an authoritative interpretation of the law" printed inside the cover. It is however authoritative in the sense that HSE would hardly take action contrary to the principles it sets out.

What do we learn? It is largely a paraphrase of the *Edwards v NCB* judgement. It says that risks must be assessed, without suggesting parameters which would help in the assessment. It uses phrases like "very low" and "very high" which are not helpful. It does however tell us one thing: a company's financial standing does not affect the issue. What is reasonably practicable for the prosperous is equally reasonably practicable for the impoverished, in law at any rate.

## "The Control of Major Hazards" – Advisory Committee on Major Hazards, Third Report (HMSO)

In its second chapter, "Living with risk", the Advisory Committee on Major Hazards gave a view on reasonable practicability which gives an interesting perspective. Some extracts are given below:

> *Paragraph 15:* We are all throughout our lives subject to risks. Some are perhaps inescapable and must therefore be accepted; others might be reduced in frequency or magnitude sometimes at the cost of eliminating compensating benefits. It is our view that the principle of balancing the efforts required to reduce risk against the possible gains has been part of our common law duty of care for generations and is certainly part of the concept of reasonably practicability in the HSWA. We think that the same approach should be followed in the field of major hazards.
>
> *Paragraph 16:* The obverse of risk is of course reliability and it is with the reliability of major hazard installations that we have been concerned. The duty laid upon employers by the HSWA requires employers to ensure, so far as is reasonably practicable, the health, safety and welfare at work of their employees. The Act also requires employers to conduct their undertakings in such a way as to ensure, so far as is reasonably practicable, that other people who may be affected are not exposed to undue risks to their health or safety.
>
> *Paragraph 17:* "Reasonably practicable" has been defined...(there then follows what has been presented in Chapter 4, namely the judgement in *Edwards v NCB).*
>
> The wording of the HSW Act is no more than a restatement of the duty of care under the Common Law which applies to us all. It is clear that in some circumstances the High Court ruling could not be controlled within financial limits which were commercially acceptable. In our 1976 report we suggested tentatively that there was a level of risk or reliability, as the case might be, which we saw as the lower limit of acceptability. It made clear also that there were circumstances and conditions where higher levels of reliability

would be both reasonable and practicable and this was a matter which would have to be judged in each individual case. The duty of care on individuals and the statutory duties laid on employers both provide for "reasonable and practicable measure", and these by the very nature of the judgement referred to above will vary with the intensity and magnitude of the risk, becoming increasingly onerous with each increase in risk; with that in mind we think there is no need to go further. It does not therefore seem feasible to us to attempt to lay upon individuals or corporations "duties" which they can not discharge. It would not be helpful to "require" a major hazard plant to be "safe".

However it is notable that the courts have interpreted this duty as being modified by what is "foreseeable" as to danger. From the point of view of major hazards the case of *Bolton v Stone* (1951) is relevant. In this case Lord Porter observed: "Nor is the remote possibility of injury occurring enough. There must be a sufficient probability to lead a reasonable man to anticipate it. The existence of some risk is an ordinary incident of life, even when all due care has been, as it must be, taken". The judgement is interesting in that the phrase "sufficient probability" is here used in a legal context.

*Paragraph 18:* The use of the word "probability" can also infer the quantification of the likelihood of given events which, with assessment of the potential consequences of the hazards, have been of particular interest and study in recent years. We deal with the subject of quantitative assessment in the next section where we consider use of quantification techniques to achieve a greater understanding of hazards and risks and hence the ways in which they can be avoided. Much has been written on the subject and it is one on which many disciplines have a contribution to make. However the results of quantification must be seen in context and treated with caution and proper understanding.

The views expressed about "sufficient probability" are significant because they point us towards quantification. The ability to measure the probability of a particular hazard making itself manifest is something to which thought should be given when considering if something is reasonably practicable.

## Training for Hazardous Occupations – a Case Study of the Fire Service HSE Occasional Paper Series OP8 (HMSO)

The majority of jobs are relatively safe and the hazards associated with them are predictable, unvarying and capable of effective control for the most part. Some jobs are not like this at all. They are inherently dangerous. The armed services, fire, police and prison services are good examples. How is reasonable practicability coped with in such occupations?

The Defence and Disciplined Services National Industry Group (referred to as NIGs in the HSE) has, as one of its tasks, the development of policy towards employment in hazardous occupations. The NIG is part of HM Factory Inspectorate.

The question of safety in training has been the major issue concerning the NIG since it was set up. The paper referred to here studied the problems.

The activities of firemen present the problem in its acutest form because the public expects and accepts that firemen and women risk their own lives to save others. So do firemen themselves: careful selection of men and women for their roles and substantial training are essential.

> The argument that follows however, is not directed to the actual operations of these emergency groups, but to their training activities when the relationship between the risks being run and the benefits that may accrue will not be so obvious. In such circumstances it is suggested, the exposure to risk may only be acceptable if those undergoing training are aware of the hazard to which they are to be exposed, fully understand why it is necessary, and appreciate the benefits which will accrue.

The study then states the conventional "reasonably practicable" qualification and points out that the judicial interpretations derive "from decisions in suits for damages arising within conventional employment". It goes on to say "it may not be appropriate to apply these criteria unmodified to training for hazardous occupations. It may, in particular, be necessary to consider the whole of an employer's activities and not an isolated incident and to have regard to the inconvenience not merely to the employer, but to the public at large in making the computation as to what is reasonably practicable".

The following elements emerge from the study.

1. It is not possible to delay a fireman's activities until the appropriate safeguards are devised. Firemen or firewomen must take rapid action in an emergency. There is always an element of risk in this.
2. Whilst some physical safeguards, such as protective clothing and breathing apparatus can be used, a fireman's main protection "must lie in a safe system of work to which his own skill and experience, the skill and experience of officers and colleagues and a high degree of discipline all contribute."
3. To achieve this "a high proportion of a fireman's working life is spent in training, retraining and exercising".
4. In training, the risks run by firemen could be minimised by measures which are perfectly practicable, eg: the fencing of a roof edge while a fireman practices rescue techniques with a stranded victim.
5. "Applied simplistically this approach would result in the additional safeguards required during training defeating the objective the training was intended to meet – an absurdity".
6. Heat and smoke, work at heights and in confined spaces generate natural fears in firemen and firewomen. Unless the fireman "has learnt to control them, there is a risk that he will get into difficulties in the hazardous circumstances of the fire ground and will himself need to be rescued. Control of fear, and confidence in himself, officers, and colleagues can be developed in training, but only if the training is realistic which may well imply exposure to risk".

How this apparent dilemma is overcome within the terms of the Health and Safety at Work, etc Act 1974 is as follows. First, it is stated that some assert that s.2 precludes the deliberate permitting of dangers during training. "This assertion can, it is submitted, only be sustained if each individual item of training is viewed in isolation. The duty imposed on employers by s.2 is to 'ensure so far as is reasonably practicable, the health and safety at work of all his employees' (NB *all*). The duty is extended by s.2 (2)(c) to include the provisions of such "instruction, training and supervision as is necessary to ensure, so far as is reasonably practicable, the health and safety at work of his employees". The time of greatest risk for a fireman or firewoman is when he or she is carrying out the prime function of a fire brigade at a fire, in a hazardous, and, initially at least, an uncontrolled environment. It is here that the greatest burden is placed upon his or her employer by s.2 (1) of the Health and Safety at Work, etc Act 1974 and he or she must provide training to discharge it. The use of the words, "all his employees" in s.2 must imply that the whole of an employer's activities are to be considered in determining what is reasonably practicable including particularly the needs of safety on the fire ground. If therefore in order to discharge his or her s.2 duty, the employer has to expose the fireman or firewoman to an element of risk on the training ground, then the Act does not prevent it, provided that all else is done by way of supervision and by provision of safeguards to ensure that overall the arrangements are as safe as they can sensibly be made.

Safeguards should not be such as to defeat the object of training. The NIG will no doubt continue the debate but for the general reader the emphasis on the role which training can play and the way the authorities have construed reasonable practicability in the special circumstances described is enlightening.

Three final references will indicate some recent attempts at reaching consensus by the Health and Safety Commission following the consultative process recommended by Robens and subsequently written into the Health and Safety At Work, etc Act 1974.

## Consultative Document – "Electricity at Work – Draft Electricity at Work Regulations 198-, Draft Approved Code of Practice and Guidance".

Our existing regulations controlling electrical safety at work have limited application to factories and some other workplaces such as construction sites. There are many places of work to which they do not apply, and in any event they were made in 1908. An up to date and universally applicable regime of statutory control is needed. As with the 1908 regulations the new proposals are couched in broad terms (as Robens recommended) supplemented by Approved Codes of Practice and guidance. The consultative document contains the following paragraph after first saying "*the notion of risk assessment, and of matching the precautions taken to the hazard posed is fundamental to the regulations".*

*Paragraph 9:* In considering what level of duty each regulation should embody it has been necessary to ask in each case whether following normal good practice would meet it. Where this is the case the regulation is expressed in absolute terms. Where, however, it is not possible for a dutyholder fully to meet the requirement, either because of the level of development of the available technology or for some other reason, the regulation is qualified, either by the introduction of the words "so far as is practicable" or "so far as is reasonably practicable" or by the introduction of terms such as "in accordance with the principles of sound modern engineering practice". The result is a structure of absolute, practicable, and reasonably practicable duties which is very similar to the structure of the duties in the existing 1908 Electricity Regulations, and which should not impose any new burdens on those who conduct their activities in relation to electricity in a safe and responsible manner.

A classic statement of the current legislative philosophy demonstrating the range of duties which can be applied.

## Control of Asbestos at Work – Draft Regulations and Draft Approved Code of Practice

Asbestos, perhaps the most widely recognised of potentially lethal materials encountered at work, domestically and in the environment, was the subject of an HSC consultative document in December 1984. The proposals were intended to implement EC directives; implement the outstanding recommendations of the Advisory Committee on Asbestos; and enable the inadequate 1969 Asbestos Regulations to be revoked and replaced by new regulations applying to all work activities.

The preamble to the document stated that the HSE gave very careful consideration to whether the duties should be absolute or qualified by the words practicable or reasonably practicable. Using the latter may seem to represent a weakening of standards thought the Health and Safety Commission so they asked HSE to justify the proposed changes. It did, and comments were invited from all interested bodies on their proposals.

Most of the draft regulations were cast in absolute terms. Four were not. These are qualified by reasonably practicable as follows.

1. Draft regulation 3(1) dealt with an employer's duty towards non-employees. HSE believes it inappropriate to impose a higher standard in regulations than that in s.3 of the Health and Safety at Work, etc Act 1974 covering the same general duty.
2. Draft regulation 13 qualified the duty to prevent the spread of contamination because total prevention was not considered practicable.
3. Draft regulation 14(2) imposed a qualified duty to provide a fixed vacuum cleaning system to allow for those cases where the expense of installing a fixed system would not be justified by any further improvement in control that a fixed system might have achieved. Whether fixed or portable the vacuum cleaning system *must* be adequate for its purpose.
4. Draft regulation 8 is left last for comment because it contained two duties, the contrast between which is most informative. The first was to prevent or minimise exposure to asbestos, and the second was to keep asbestos exposure below the control limit. An absolute duty to prevent exposure would be a high cost exercise, and some asbestos processes would become uneconomic, be discontinued and incur social costs accordingly. These social costs would be incurred even though the risks to employees might be very low – hence HSE suggested so far as is reasonably practicable was the appropriate standard in the circumstances. The second duty was absolute. Regardless of cost, the asbestos exposure must *not* exceed the control limit. HSC believed that with most duties being made absolute, plus HSE's declared intentions of enforcing standards stringently, better standards and better employee protection would result.

In the event, regulation 3(1) of the Control of Asbestos at Work Regulations 1987 retained the qualification so far as is reasonably practicable in relation to the duty of employers towards non-employees as proposed in the draft.

Draft regulation 13 appeared renumbered in the regulations and when passed as regulation 12 retained the so far as is reasonably practicable qualification.

In the regulation requiring a vacuum cleaning system, it was said a fixed system had to be provided if it was reasonably practicable to do so.

Regulation 8 also appeared, as proposed, with the primary duty to prevent exposure but where prevention was not reasonably practicable, to reduce to the lowest level that was reasonably practicable by means other than by respiratory protective equipment.

Where exposure reduction is not achieved to below the control limits, then as well as the steps outlined above, the employer has to provide suitable respiratory protective equipment to reduce the concentration of asbestos in the air inhaled by the employee to a concentration which is below those control limits.

## Draft Prevention of Damage to Hearing from Noise at work Regulations 198-

First the subject of a United Kingdom Consultative Document in 1981, the control of the risk from noise at work was then the subject of an EC directive finally adopted in May 1986. This has a deadline for implementation set by the European Community for 1.1.90, and the United Kingdom regulations have been promised in good time for this.

A second set of draft proposals was published at the end of 1987 to which responses were made by interested parties until 30.6.88. The primary duty in the proposed regulations is that every employer will have to reduce *the risk of damage* to the hearing of his or her employees from exposure to noise to the lowest level reasonably practicable.

The employer's second duty is to reduce *noise exposure* of employees so far as is reasonably practicable other than by the provision of personal ear protectors. This wording precludes the easier and often cheaper option of wholesale indiscriminate provision of personal ear protection as an alternative to dealing adequately with the source of the noise itself.

In relation to the provision of ear protection, however, when recourse to this provision is legitimate, the duty is not qualified by the phrase reasonably practicable. *All reasonable steps* have to be taken to provide suitable and efficient personal ear protection. *All reasonable steps* also have to be taken to see that proper use is made of what is provided and that it is kept efficient and in good repair. The extent of the employer's efforts to demarcate ear protection zones and to see to it that none of his or her employees enter such zones unless they are wearing ear protectors is, however, qualified by the reasonable practicability clause. In the event, the phrase "all reasonable steps" was not used in the Noise at Work Regulations 1989. The phrase used was "so far as practicable" which eliminates considerations of cost (see page 39).

## The Control of Substances Hazardous to Health Regulations 1988

Reference was made later to these regulations when officially published standards of performance (MELs and OESs) and their standing in relation to reasonable practicability issues was explored. The 1988 regulations, which will be the governing statutory instrument on matters of occupational health well into the next century, use the qualifying phrase so far as is reasonably practicable four times. On the first occasion it governs the extent of the duty of employers to protect those who are not their employees (regulation 3).

The other three times when employers' duties are qualified are all in relation to the prevention and adequate control of exposure to substances hazardous to health (regulation 7). What is actually hazardous to health is the subject of a comprehensive definition forming part of the regulations themselves (regulation 2).

Under the regulations, the primary duty is always to prevent the exposure of employees to substances hazardous to their health. Only where prevention is not reasonably practicable, can the employer pursue the second option of exercising adequate control for employees' protection. The next use of the reasonably practicable criterion is where employers are required to prevent or adequately control exposure by means other than by the provision of personal protective equipment.

The reasoning behind this lies in the inherent weakness and unreliability of human behaviour compared with engineering methods of control, such as enclosure or the use of well-designed and efficient local exhaust ventilation equipment. The last use of the qualification in regulation 7 was that described earlier in relation to the achievement of a level of exposure *as low as is reasonably practicable and in any case below the Maximum Exposure Level.* The likely interpretation of the phrase as soon as is reasonably practicable used in regulation 7(5)(b) to determine the urgency of action of an employer discovering that an Occupational Exposure Standard (OES) had been exceeded, is, it is submitted, as follows. The higher the degree of hazard to employees from the substance concerned (usually associated with a low figure for the relevant OES) the quicker would remedial action be expected.

## Consensus

The subject of "reasonable practicability" and the standards implicit in its operation as part of the legal structure, are reflected in consultative documents which now precede any new legislation. Consultative documents and what follows them in legislative form are products of consensus. It is therefore extremely important for any employer or employee to realise that he or she can play some part, however small, in the law-making process. The former will work through his or her trade association and the CBI, and the latter through his or her trade union and the TUC. Both these bodies have three members each on the Health and Safety Commission and a similar representation on Advisory Committees at national level, and also on Industry Advisory Committees (IACs) referred to earlier.

# Chapter 5

## Sources of official guidance

Although the phrase "so far as is reasonably practicable" may be relevant both in the courts after accidents and mishaps and in workplaces when precautionary measures are being devised, there is guidance in existence which indicates standards reflecting what will be accepted as reasonably practicable. It is always open to a court to disagree in a particular case but, in general, following established guidance from reputable sources will meet most eventualities.

## Information

So far this book has explained how the phrase so far as is reasonably practicable is used in our occupational health and safety law and the reasons behind its use. It has also described the test established by the courts for determining whether or not all that is reasonably practicable has been done in a particular case. The same test is relevant whether the scene is the High Court or the workplace. There is however an enormous difference in the ease with which the computation can be made.

In a court the subject of reasonable practicability nearly always arises as an issue after an incident involving injury or disease. When it does, the odds are stacked against the defendant or accused because the evidence will almost invariably tend to suggest that there were further simple and straightforward measures he or she could have taken, some of which would have been reasonably practicable. The courts, and it is the same for enforcing authorities investigating incidents, will have the wisdom of hindsight. What the person who has the duty to fulfil in practice needs, is foresight.

How can employers be sure they are doing all that they need to? The next seven headings will take them a fair way along the road towards a satisfactory answer to this question.

### Compliance with absolute requirements

To be seen to be taking the law seriously by meeting all its absolute requirements (ie when they are precise and unqualified) must be the first thing for anyone to do. This must be so, if for no other reason than that non-compliance is a punishable offence. We saw earlier than not only can fines be imposed upon companies but also upon certain officers of the company and upon employees. There is now a statutory requirement to have a written company policy statement on health and safety which has to show what the organisation and arrangements are to implement it. The individual manager is under the spotlight as never before, and his or her accountability is more easily subject to examination and question.

Occasionally the courts find an individual guilty of an offence where he or she is alleged to have failed to fulfil a duty, and that person is then subject to a penalty, provided that the prosecution proves beyond reasonable doubt what it has alleged.

## Compliance with Approved Codes of Practice (ACOPs)

Earlier it was stated that because general duties taken from the common law duty of care (now embodied in s.2 of the Health and Safety at Work, etc Act 1974 were expressed in such very broad terms there was need for practical guidance. The concept of Approved Codes of Practice (ACOPs) was established and where they exist they indicate an acceptable standard to follow.

Important as the freedom of an individual to choose methods of compliance other than those in an ACOP may be, it has to be emphasised that if enforcement action follows some failure resulting in an incident involving personal injury, it will never be easy to argue successfully that following a method other than the one in the ACOP has been satisfactory. The evidence will all be suggesting that the employer was wrong and that following the ACOP would have been better. There is no guarantee that it would have been better but the evidence will certainly not easily persuade those sitting in judgement that the employer's method was better in view of the evidence about what actually happened and what more could have been done.

## Following analogous statutory standards

Sometimes, because of the restrictions of their enabling statutes, regulations only apply in certain limited classes of premises, eg regulations made under the Factories Act 1961 only apply to factories as defined in s.175 of that Act, or as allowed for in ss.123 – 127 (relating to electrical stations, certain institutions, docks, wharves, quays, ships, building operations, and works of engineering construction). Nevertheless, the identical hazards which they are intended to eliminate or control exist in other workplaces not covered. Although the other premises will be subject to the general duty on the employer to do what is reasonably practicable under s.2 of the 1974 Act, no detailed specific regulations exist. These anomalies will disappear in the course of time. Meanwhile, the strictly non-applicable regulations should not be ignored as sources of guidance merely on the narrow, technical basis of their non-applicability.

Two examples of regulations falling within this category relate to the protection of eyes and the hazards involved with highly flammable liquids. In both these cases the regulations were made under powers contained in the Factories Act 1961 just before the Health and Safety at Work, etc Act 1974 came into effect. No workplace outside the definition of the term "factory" in s.175 of the Factories Act 1961 can be subject to either of the detailed sets of regulations. Nonetheless, there are many workplaces where eyes are at risk and where highly flammable liquids are a potential source of danger to those who work there. Wherever there may be relevant regulations, their standards should be followed. As suggested above they should never be ignored merely because they do not strictly apply to areas where the hazards they deal with exist.

## Following non-statutory standards

In September 1983 the Health and Safety Commission made a statement on its future policy towards reference to standards. In 1984 it published a booklet called "Standards significant to health and safety at work". The booklet, out of print when the edition was written, has since reappeared and is available from HMSO (ISBN 0 11 885496 8).

The listing of standards "creates no obligation of a legal character on either manufacturers or users. It represents no more than a statement of the fact that a particular standard is used by the HSE, which has found it particularly helpful in giving advice and guidance on those particular aspects of safety that the standard covers". This is a predictably guarded but quite significant statement. Later, under the heading "Product Standards" we find the statement "HSE inspectors have not prosecuted for a breach of statutory duty under s.6(1) (of HSWA) where an article has been shown to have been manufactured in conformity with a reputable and relevant current standard which adequately and satisfactorily covers safety aspects and is properly applied". Six current British Standards have been formally approved by the Health and Safety Commission and have the status of ACOPs described earlier.

In the HSC statement we also find an acknowledgement that HSC and HSE "recognise a duty to all concerned to be as clear as possible as to the expectations they have for the performance of these duties, and intend to use and refer to British (and other) standards wherever these can make the appropriate contribution". Steps have been taken to see that HSE will participate in BSI Technical Committees "on the basis that it expects to use the resulting standards and regards it as indicative of its safety and health requirements".

Standards may be mandatory if applied by regulation. They may also be approved as ACOPs by the HSC and have the effect described earlier; or they may be referred to in other guidance from HSE, written in support of regulations and ACOPs. "Guidance does not create legal obligations nor can it always deal fully with all the relevant safety implications of a particular situation. It does however define the attitude of the regulatory authority and thereby assists both the courts and individuals in considering what ought to be done."

Since prosecution for offences under our occupational health and safety law may only be instituted by inspectors, or with the permission of the Director of Public Prosecutions (s.38 of the Health and Safety at Work, etc Act 1974), and because civil proceedings under ss.2 – 7 of the Act are debarred (s.47), it is unlikely that any action will be taken contrary to that stated as the authority's line. Reference to standards in guidance will continue to be the system most frequently employed by HSE to draw attention to standards and encourage their use. HSC has stated "We intend to pursue a deliberate policy where possible, of stating in guidance, eg in connection with new regulations, what the relevant standards are, with an indication, if necessary, of the extent to which they can be relied upon as a guide to the underlying legal requirements".

## HSE guidance

As stated already, HSE guidance has no legal status per se when compared either with regulations made by a Minister of the Crown under powers given by the Health and Safety at Work, etc Act 1974, or with Approved Codes of Practice having the formal stamp of approval from HSC (which, incidentally, also requires Ministerial consent and prior consultation with interested parties).

HSE Guidance Notes are published in five categories. In each category a Guidance Note has a serial number which is preceded by the initial letters indicating the category as follows: General, GS; Chemical Safety, CS; Plant and Machinery, PM; Medical, MS; and Environmental Hygiene, EH.

Three questions arise in connection with non-statutory guidance emanating from HSE. What happens if employers do not follow it? What happens if they do? How do they know of the existence of any particular guidance?

If employers do not follow HSE published guidance on a particular matter, they may face an allegation of non-compliance with s.2 of the 1974 Act. The allegation may in whole or in part be supported in court by evidence given probably by the author of the guidance itself, or a specialist colleague, as an expert witness. The defendants may argue that they chose to achieve the law's objective by a different route. The chances are (as with not following an Approved Code of Practice) that the reason they are in court is that this route ended in failure and it is as a consequence of the failure that the allegation was made that they did not do all that was reasonably practicable.

If employers do follow HSE published guidance in a particular area and things still go wrong for any reason, as they may well do, the fact that an officially prescribed procedure or any other thing recommended in guidance was followed is bound to be very strongly persuasive that they were doing all that is officially considered reasonably practicable. A court may not agree (a court is always the final arbiter), but it is going to be a rare event for HSE to lay charges against anyone following its own guidance, so the chance of such an issue getting into court at all is very small.

How do employers know what guidance exists, and that their own collection of relevant material is up to date? The section entitled Information Sources (see pages 91-92) sets out some of the sources which may be used to help in keeping employers adequately informed and up to date. It is critically important that every employer has a system for keeping information sources up to date. If this is not done there will come a time when the "current state of knowledge and invention" has moved forward and left such employers behind with yesterday's outdated and unacceptable standards and knowledge.

## Voluntary industry codes

Some industries have produced their own voluntary codes for their members. Take for example "A Code of Safe Practice at Fairs" where both the safety of employees and the public may be affected if anything goes wrong. Its foreword reads as follows:

> Although this code does not of itself have the force of either law or an Approved Code of Practice, the HSE has instructed its inspectors to take it into account when considering whether there is compliance with statutory requirements. Failure to follow the guidance or to provide equally effective measures may lead to action by inspectors ranging from advice and warnings to the issue of enforcement notices or even prosecutions, although ultimately it is for a court or tribunal to decide whether there has been compliance with the law.

The code was prepared as a result of a joint initiative by the HSE and the three main organisations in the amusements industry. The paragraph above indicates its status, and hence the chances are minimal of anyone sustaining an argument successfully that standards falling short of those in the code are good enough.

## Custom and practice in the trade

If an employer's standards are below those of his or her peers in a particular area of industry, an accusation of not doing all that is reasonably practicable has a good chance of succeeding if tested in the courts.

In these days of Industry Advisory Committees (IACs), described appropriately elsewhere as "mini commissions" because of their tripartite composition (CBI/TUC/HSE), the particular occupational health and safety problems of an industry are identified. Thereafter industry working parties and subcommittees produce relevant guidance for their own industry's members. Such groups are serviced by HSE's Factory Inspectorate National Interest Groups (NIGs, formerly referred to as National Industry Groups) who specialise in their own industry's major problems, and in working out practical solutions to them. This is done with the very close collaboration of those engaged in the industry itself. These arrangements produce valuable guidance on what is feasible and this becomes accepted and regarded as "custom and practice". Not all industries are yet equally well served but pragmatic indications of what has actually been found to be feasible and practicable across a whole spectrum of a specific industry have started to appear as a result of the IAC system as commented earlier. Inspectors, now more deeply involved in the industries' problems than was the case a decade ago, tend, not unnaturally, to be rather sceptical of any they meet in their industry who try to tell them that certain standards are neither achievable nor reasonably practicable.

Readers will appreciate a caveat publicised during the "Site Safe '83" campaign run by CONIAC (the Construction Industry Advisory Committee). It was a reminder by the Health and Safety Executive that if a bad practice was commonly regarded as custom and practice in the trade (a classic example was the lack of adequate protection afforded to steel erectors), the HSE would not be very sympathetic towards anyone attempting to perpetuate the bad old habits, hallowed by long use but still bad. Leaders in the industry are demonstrating that good standards are possible and feasible with careful design and planning. It is these standards that will be looked on as custom and practice. What is achieved by the industry leaders will steadily extend the boundaries of what is regarded as within the state of "current knowledge and invention".

# Chapter 6

## Management action plan

The health and safety of staff has to be planned along with every other management activity. It is helpful if the planning process is broken down into stages which are presented in a logical way.

Health and safety issues are not resolved automatically. Hazards are not like flowers that wilt and die away if you ignore them. Quite the reverse is, in fact, the case. Any risks from the hazards present in the workplace grow in seriousness in direct proportion to the neglect that is shown towards them. But enough of analogy. This chapter is meant to serve as a recipe to determine what is reasonably practicable in health and safety in the workplace and to guide those having to make decisions and judgements on the subject about all the factors they should be considering.

## Management action plan

This action plan suggests seven positive steps which can be taken by management to assure itself that it is complying fully with its statutory duties and is in a position to justify and defend its decisions and judgements if called upon to do so.

### Step 1 Identify hazards

The compilation of a comprehensive hazard inventory is the essential starting point in the effective management of occupational health and safety. All possible sources should be utilised to create a fund of information. Employers should not be worried about listing obvious dangers. The more obvious the danger the more likely that it will have been covered by specific statutory regulation with which they are already complying.

Employers are only concerned with foreseeable risks so an intimate knowledge of their own unique past experience is a good starting point for identifying hazards. Other sources of information, for example, are the data sheets required to be provided by manufacturers and suppliers under s.6 of the Health and Safety at Work, etc Act 1974, as amended by s.36 and schedule 3 of the Consumer Protection Act 1987 and the labels put onto dangerous substances under the Classification Packaging and Labelling Regulations 1984.

As many sources as possible should be consulted and in areas of uncertainty about either hazards themselves, or the degree and nature of the risks they pose, specialist advice should be sought. The law's view of reasonable people is that they know the limits of their competence and seek expert help when they stray beyond them.

### Step 2 Match all identified hazards with the appropriate absolute statutory requirements

This step will ensure that in cases where the law gives no leeway it is possible to check that standards of compliance are satisfactory. Any plant which requires to be "properly maintained" falls into this category, and covers such items as pressure vessels, lifting machines, lifting gear, hoists, etc. Similarly a machine classed as dangerous *must* be securely fenced.

### Step 3 Match all identified hazards with any other authorative guidance

Earlier we referred to a range of acceptable guidance. Identified hazards should be matched with all appropriate items in the range. By this stage the employer will have discovered that the hazards index has very few items remaining on it where there is no indication at all of the standard of performance expected from him or her.

### Step 4 Isolate residual hazards where no external regulatory control or guidance exists

Only at this stage need employer's even consider the risk/sacrifice equation in the *Edwards v NCB* judgement *as a starting point*. They should look at each hazard in this category and try to establish some incident experience from their own history. No incidents should be left out because they are rare. Both the *Edwards* and *Marshall* cases involved rare events which ended in fatalities. The experience of others should be sought through IACs, NIGs, trade associations and other relevant professional bodies. Step 4 concerns itself particularly with the known and/or potential frequency of hazardous events. The next step is concerned with the severity of the consequences of a hazard.

Severity of consequences is a crucial aspect of the *Edwards v NCB* test.

### Step 5 Compile a hazard severity index

Each hazardous activity should be considered in conjunction with the worst consequences which can arise from it. Can these include fatal or major injuries or life-shortening disease? How many people could be affected? If employers construct a rough and ready scale from their estimates they will know that an activity which could cause fatal injuries to several people will require far more attention than an activity which could only cause a minor injury to one person. The very least which can be derived from this review is a list of priorities. The test for determining what is reasonably practicable refers to *gross disproportion* between risk and the cost of averting it. Precise mathematical calculations are never possible but Lord Reid's words in the *Marshall* case are relevant. "As men's lives may be at stake, it should not lightly be held that to take a practicable (ie physically possible) precaution is unreasonable". If an employer's researches have shown even a patchy history of severe consequences he or she should know that the control regime needs to be very strict and effective to be within the law. A refined system of looking at frequency and severity rating is contained in Chapter 8.

The building up of an incident database can be an invaluable source of information to influence the direction of management action in the prevention of similar mishaps. It should not confine itself to injury accidents but also include damage-only incidents and "near misses".

### Step 6 Establish an appropriate prevention policy and control measures

Once identified hazards have been sized up, a regime is needed to ensure that performance will be up to standard (the many types and sources of these have been described). Inform, instruct, train and retrain everyone to the highest level. Allot responsibilities for achieving objectives and make those concerned accountable. Embody objectives in the policy statement; involve safety representatives to the full at all stages; and prepare emergency and contingency plans to cope with the mishaps which cannot be avoided.

### Step 7 Establish effective monitoring

However good an employer's performance in Steps 1 – 6 are, he or she needs to instigate a system of thorough monitoring. Again the involvement of safety representatives and safety specialists is highly relevant, as is the keeping of accurate and easily understood records of what monitoring has revealed. The final step is for everyone in management to have a procedure for seeing that the omissions and weaknesses discovered during monitoring are avoided in the future by the implementation of effective remedial measures. The first task in any monitoring exercise is to check that previous monitoring has borne fruit. There is a fuller description of monitoring in Chapter 9.

## Complete management health and safety strategy summary

1. Compilation of hazard index (using list of hazardous agencies).
2. Application of frequency and severity rating numbers.
3. Review of priority allocation for actions.
4. Setting acceptable standards.
5. Continuous collection of relevant data on incidents/accidents.
6. Establishing an effective monitoring/auditing procedure.
7. Implementing routine senior management reviews of the working of items 1 – 6.

# Chapter 7

## Hazard identification

Some hazards such as falls, fires and high voltage electricity, etc are obvious. Many other hazards such as oxygen deficiency, carbon monoxide and pathogens are not obvious — they are very insidious.

The essential starting point for all thorough and successful health and safety management must be the undertaking of a review of all the possible threats that could be encountered. No realistic management of hazards is possible without such a review and the hazard index which arises from it.

The compilation of a comprehensive hazard inventory or list is the essential starting point in the effective management of occupational health and safety. Without one management do not know what they may be called upon to manage, or what hazards they may have to contend with. Therefore they cannot plan effectively to minimise the risk of injuries to people or damage to property from such hazards. Moreover, they will be in no position to say that they have done all that is reasonably practicable to comply with the law. How can management, for example, establish and maintain a safe system of work when they do not know all the components which may need to be considered?

## Agencies causing injuries to people

Although the chapter presents a comprehensive list of the major categories of agents responsible for causing injuries to people, it does not pretend to be complete because no such list ever could be. New dangers emerge as technology advances. Whereas the law of gravity plays a significant part in many of today's accidents, in a future workshop in space it would not be a problem. Nevertheless we would then have to contend with weightlessness and any attendant risks that it might present. We would be exchanging one hazard for another.

## Hazard and risk

In the last paragraph the words hazard and risk were both used. It is important that the difference between them is appreciated. A hazard is a physical situation with a potential for harm to life and limb and a risk is the probability that a hazard may be realised in a given span of time or the probability that an individual may suffer injury as a result of the realisation of a hazard.

In relation to hazardous substances the Health and Safety Executive's leaflet "Hazard and Risk", produced as part of its guidance for the implementation of the Control of Substances Hazardous to Health Regulations 1988 (COSHH), puts the distinction as follows: "The hazard presented by a substance is its potential to cause harm. It may be able to make you cough, damage your liver or even kill you. Some substances can harm you in several different ways eg: if you breathe them in, swallow them or get them on your skin". It continues "The risk from a substance is the likelihood that it will harm you in the actual circumstances of use". The guidance leaflet goes on to say that "Poor control can create a substantial risk even from a substance with a low hazard. But with proper precautions the risk of being harmed by even the most hazardous substance can be adequately controlled".

In other words, you can lessen the risk from a hazard by taking proper precautions but the hazard is still there. If you lessen your vigilance the risk from the hazard will increase. With substances there are regulations dealing with labelling and apart from the visual symbols indicating flammability or toxicity, there are phrases identifying the nature of the hazard such as "toxic by inhalation" or "irritating to the eyes". Labels are also required to show other phrases giving indications about how the risk can be minimised, such as "do not breathe dust", "wear suitable gloves", or "avoid contact with eyes".

## Hazardous agencies

Eighteen categories of hazards can be listed which between them cover most of those likely to be encountered under normal circumstances at work. The object of listing and describing the categories is not to achieve an all-embracing, scientifically accurate, one-hundred-per-cent complete list but to make the reader aware of the range of possibilities which may have to be faced in the workplace in question.

*1. Gravity, loss of balance*
As human beings move about undertaking various tasks they are prone to lose their balance and fall. In the process of falling, whether from a height or on the level ground, they can sustain injuries.

As a general rule the greater the height from which a body falls, the greater the injury sustained. Whilst this is the general rule there are individual cases where it does not operate. Some people may fall from a considerable height and only be slightly hurt. Others may fall only a few feet and sustain fatal injuries.

The hazard of falling is a universal one and a great deal of management effort has to be expended to ensure that the risk is kept at an acceptable level. Consider some of the activities associated with the prevention of falls. They include the provision of suitable floor coverings kept clean and free from obstructions, slippery substances, trailing cables, etc. Also the provision of handrails on staircases, guard-rails and toe-boards on elevated walkways, catwalks, balcony edges, etc. They include the provision of suitable ladders, safe access on to vehicles by the provision of handholds and footholds and so on. There are many other examples which could be given but even if they were all catalogued and dealt with adequately, people would still bump into a desk and stumble; move awkwardly as they carried a box, lose their balance and fall; or simply turn a corner in a corridor and collide with someone else coming the other way in a hurry with the result that both people fall.

It will have been concluded by now that, in some of the cases given under this heading, the law must have set down some rules and standards to prevent falling accidents. It has done so.

For example, an employer does not have to think about the hazard of falling downstairs and decide whether it would be desirable for there to be a handrail at the side of the staircase for people to hold on to. In office premises, for example, s.16 of the Offices, Shops and Railway Premises Act 1963 covers the point about handrails. They are legally required to be there. It also covers floor construction and maintenance. Where there is no specific mention of the provision of equipment or facilities to prevent falls, such provision by the employer is implicit in s.2 of the Health and Safety at Work, etc Act 1974. There it is required that measures be taken "so far as is reasonably practicable as regards any place of work under the employer's control, the maintenance of it in a condition that is safe and without risks to health and the provision and maintenance of means of access to and egress from it that are safe and without such risks".

*2. Falling or projected objects and projectiles*

It is not only people who fall but also objects of all sizes, unless they are prevented from doing so. Some of these fall on to people and injure them. So far this category of hazard is identical with the last one, with the force of gravity applying to inanimate objects and not to people, who are merely the unsuspecting targets. As well as the force of gravity playing a part as described above, there is also the case of objects which obtain a momentum from some source or other and become projectiles which hit people and injure them. Eventually of course the momentum is spent and the projectile surrenders to the force of gravity and falls to the floor.

There is no need to give a long list of examples of the first class of falling objects — the can of paint from the place on the ladder where it slipped from the painter's grasp; the loose item falling from a poorly slung load carried by a crane; a box knocked off the top of a stack when the fork lift truck collided with it — everyone can envisage their own examples.

As for the other class, the projectiles, typical examples are: the particles of metal and stone thrown off from the periphery of a quickly rotating abrasive wheel as a chisel is being sharpened or the chips of wood flying off a high speed cutter on a routing machine.

*3. Manual handling*

Wherever loads are handled, and this literally means at every workplace, there is a risk of injury to those involved. The load may be too heavy to lift safely, too hot and causes burns, have sharp edges which inflict cuts, or it could be contaminated with harmful chemicals which affect the exposed skin and so on.

The hazards associated with manual handling of loads present management with a challenge. There are clearly certain loads which are so light that injury from them is highly unlikely or even physically impossible. There are, equally clearly, certain loads which are so heavy, big or cumbersome that injury is virtually certain if anyone attempts to lift them unaided. In between the two extremes there lies a whole range of loads where there is a possibility of injury during manual handling depending on the stamina, strength, age and degree of experience of the persons doing the handling. Furthermore, a number of other variable factors including the location of the operation, the style of supervision and the number of times the handling operations are carried out, could all influence the risk.

*4. Hand tools*

Everyone who has wielded a hand tool at work (a hand tool is any tool which can be carried easily by the person using it) knows that they can inflict damage. Who has not hit his or her thumb with a hammer despite aiming to hit the nail? Who has not skinned his or her knuckles when the spanner slipped? Who has not let the paring knife slip when preparing vegetables and cut a finger? In practically every workplace some type of hand tool is used. In most instances, where not in skilled and practised hands, we all know they can bang, cut, scratch or abrade their users with varying degrees of severity of outcome.

*5. Moving parts of machinery, plant and apparatus*
Since the days of the industrial revolution the dangers associated with moving machinery parts have been recognised. As long ago as 1844 Parliament tackled the dangers by bringing in an Act to ensure the provision of "secure fencing", as it was called, for such things as mill gearing and transmission machinery which festooned the original textile mills and in which workers often got caught up or entangled with grievous results.

More recently the British Standards Institution has brought up to date its "Code of Practice for Safety of Machinery" (BS5304:1988) in which all the classic machinery hazards are identified very graphically. Moreover, after illustrating the range of machine parts capable of causing accidents by crushing, trapping, cutting, nipping, dragging in and entangling, etc, the British Standard devotes its pages to imparting a wealth of illustrated methods of suitable guarding to prevent the type of accidents which still comprise a significant proportion of the total.

*6. Vehicles (track and free moving)*
Hazards connected with vehicles are principally related to their ordinary motion as they are driven from A to B. Pedestrians are hit and sustain injuries or are killed in exactly the same ways as would occur on the public highway.

However, in and around premises the dangers are frequently made worse by poor visibility, obstructions, poor design of the roadway system and by the unavoidable need to reverse the vehicle to gain access to loading points or bays. In a reversing vehicle the driver's vision is notoriously bad and a very common and worrying time for vehicle accidents is during this manoeuvre. Not infrequently the victims of the accidents are the banksmen who should be in control of the reversing. They pass from sight and are run into. Other dangers are associated with loading, unloading and the maintenance of the vehicle.

Again, with vehicles running on a fixed track, trapping of people between the vehicles and fixed structures is a quite common source of accidents. There is also the problem of limited visibility from the driver's cab.

Whilst dealing with the general question of vehicles and the dangers they pose, it is worth dwelling on the fact that for many who travel extensively, such as sales representatives, a very large part of their working time may be spent on the public roads. This means that their exposure to hazards from other drivers, who are not amenable to control or influence, is quite a challenge. Some employers deal with the hazard by concentrating upon teaching their driving staff the skills of defensive driving.

*7. Fire*

Fire is a ubiquitous hazard and there are no premises which are immune from its effects. There are very many different sources of ignition, from the common ones associated with electricity, arson, careless smoking habits and open flames used in processes, to the less common ones of static electricity sparks generated by the flow of certain liquids through pipes, friction, spontaneous combustion and lightning. It is wise to assume that there will *always* be a source of ignition in every workplace.

Two significant factors are worth mentioning in connection with the hazard of fire. The first is that the great majority of fires start in a very small way and hence can be dealt with most easily in their early stages. The second is that people are seldom affected by the flames of a fire. They are overcome first by the toxic and suffocating fumes and hot gases which the fire gives off in great abundance. By the time the flames reach most people they are already dead from asphyxiation.

*8. Explosions (pressure, chemical, dust)*

Three main types of explosion are envisaged in this category. The first is where a vessel containing a gas (or a mixture of liquid and gas) no longer has the strength to contain it. The pressure proves too great and the vessel ruptures with the violent expulsion of its contents and the widespread projection of both large and small fragments of the disintegrated vessel. The second is an explosion of a chemical nature where a reaction takes place with a very sudden increase in pressure and temperature. This is the violent type of explosion which we associate with the rupture of the vessel concerned and a fireball which engulfs the surrounding area. It is akin to a military-style detonation. The third type of explosion is that which takes place under certain conditions where a dust cloud of fine particles is ignited and creates a pressure wave within a building. The primary explosion disturbs more dust from within the building creating another dust cloud, which in turn is ignited by the smouldering of the fires remaining from the first. It is almost always the secondary explosion which wrecks the building.

*9. Chemical assault*

Practically every substance encountered in the workplace has a potential for causing some sort of harm. It all depends, as Paracelsus the fifteenth century scientist first said, upon the dose. The truth of his saying is best illustrated in the field of drugs. Under the prescribed dose drugs can effect a cure but an overdose can kill.

The body can be assaulted by chemicals in many ways. The principal mode of entry by toxic materials is inhalation. Airborne particles, liquid droplets, fumes or gases are breathed in and are then transmitted via the lungs to the blood and further distributed throughout the body to those organs for which they have a particular affinity. Lead and mercury have an affinity for the brain, cadmium for the nerves and benzene for the bone marrow.

Apart from this entry route, elaborated upon further in the next section, chemicals can effect the eyes and the skin and can also be ingested via the mouth to the gut.

Over the last year or so the role of the skin, broken or unbroken, as an entry route of toxic materials into the body has been under scrutiny. It is now believed that more prominence should be given in the future to the role of the skin as an entry route.

Further, the skin can itself be attacked directly and damaged by a range of substances which degrease it, sensitise it, burn it, or otherwise cause irritation and the commonest of all skin conditions – dermatitis.

Fortunately these days, under the regime of legislation requiring labels, we are able to discover by looking at a symbol what type of harm attaches to exposure to the contents in the container. We know at once whether it is very toxic, toxic, corrosive, harmful or irritant. Furthermore, suppliers and manufacturers of substances for use at work are under a statutory duty in s.6 of the Health and Safety at Work, etc Act 1974 as amended by the Consumer Protection Act 1987) to provide data sheets giving information on any health or safety risks associated with their materials.

*10. Contaminated atmosphere (includes oxygen deficiency)*
As indicated in the last section, inhalation is the commonest mode of entry into the body for toxic substances. This being the case, there is, not surprisingly, official guidance on Occupational Exposure Limits (ie: what is considered to be an acceptable limit to which the average worker can be exposed daily without suffering any adverse effects). Such limits need to be carefully studied. The Control of Substances Hazardous to Health Regulations 1988 require exposure to substances harmful to health to be prevented if it is reasonably practicable to do so. If it is not reasonably practicable, then adequate control has to be exercised. Control is only deemed to be adequate when exposure is reduced so far as is reasonably practicable and *in any case below the appropriate Maximum Exposure Limit or Occupational Exposure Standard for the substance concerned, as published in the current Health and Safety Executive Guidance Note* described further below.

In the Health and Safety Executive's Guidance Note EH40/89 all the current limits are given — 29 in the case of Maximum Exposure Limits (MELs) and several hundred in the case of Occupational Exposure Standards (OESs). The Guidance Note also indicates substances for which changes are proposed and which ones are under review by the Health and Safety Commission's Advisory Committee on Toxic Substances and WATCH (the Working group on the Assessment of Toxic Chemicals).

It will be appreciated that although there are a few hundred substances with officially recognised limits, there are many thousands of others in use in industry where there is as yet no such guidance. Here the users have to devise their own control strategies and measures to standards which they themselves have to be in a position to justify as being stringent and adequate enough for the substances concerned, and for the protection of the workers exposed to them. It is to check the efficacy of control measures under such circumstances that health surveillance is so important.

This is a good example of where being confident of having done all that is reasonably practicable means a great deal of background work and being able to show in writing, in the formal assessment of health hazards required to be undertaken by The Control of Substances Hazardous to Health Regulations 1988, that hazards are both understood and under control.

As well as contaminants in the workplace atmosphere causing harmful effects, the insidious nature of oxygen deficiency has to be recognised. Enclosed or other confined working spaces can, for a variety of reasons, not have enough oxygen to sustain life. Such potential death traps need to be known about and the appropriate procedures used when entry to them is effected.

*11. Biological assault*

Many workplaces expose those who work in them to harmful bacteria, fungi, spores and micro-organisms. In some the harmful agencies are in plant such as humidifiers, air-conditioning equipment or in cutting oils and coolants used in the engineering industry. Humidifier fever and Legionnaires' disease are well known. Outbreaks of the latter have brought about the deaths of a number of people, some of whom did not work in the premises where the outbreak originated.

In other workplaces the harmful agencies are to be found in the general environment. In this category fall the self-descriptive conditions of farmers' lung and mushroom growers' lung. In the cotton textile industry cotton dust itself is the cause of a condition known as byssinosis.

Sewer workers are exposed to Weil's disease or leptospirosis, which is caused by a bacterium to be found in water containing infected rats' urine.

Those employed in the health care sector are always at risk from the infections contracted by those in their care. Because of this, an important part of the health surveillance regime under The Control of Substances Hazardous to Health Regulations 1988 includes the establishment of the immunological status of the exposed employees. The same applies to those who work with pathogens in laboratories.

Workers with animals, whether in agriculture, the veterinary services or abattoirs, are exposed to glanders or brucellosis, diseases of cattle which can be transmitted to man.

Workers in tanneries and those handling certain hides and

skins from other parts of the world where anthrax is still to be found, can pick up the spores and become infected by them. With these few examples taken almost at random, it can be seen that the spectrum of risk in this particular category is very broad.

*12. Animal assault (including human)*

Farm workers and anyone else handling animals (eg: in zoos, safari parks, laboratories) can be subject to animal assault. Bites, scratches and crushing injuries are to be expected where wild animals and certain domestic cattle are kept confined and have to be tended, fed and kept clean by their human keepers. The human species is part of the animal kingdom and some of its members work in places where other members become physically violent towards them.

Those most at risk, according to a recent study, are those who handle money (eg: in banks, shops, building societies); the caring services (eg: nurses, teachers, home-helps, social workers); those who have inspection and enforcement duties (eg: traffic wardens, park keepers); those who work with known potentially violent people (eg: prison officers, landlords, policemen); and those who work alone (eg: home visitors, taxi drivers, domestic repair workers).

Actual physical abuse, or the constant threat of it, is both a physically damaging hazard, in the case of the former, or a stress-producing psychological one in the case of the latter.

*13. Drowning*

The hazard of drowning needs very little by way of elaboration. People who earn their living at sea, on barges, or anywhere else in association with water at harbours, docks, quays or sports facilities run some risk of getting into the water and being unable to get out unaided and drowning in consequence.

*14. Natural phenomena (storm, wind, ice, fog, heat, cold, lightning)*

In any working conditions or environment the hazards already there can be influenced by the forces of nature. The influence is very seldom for the better. Storms and high winds make access more precarious and make certain work at heights impossible or very dangerous indeed (eg: roof sheeters, steeplejacks, linesmen in the electricity supply industry).

Fog makes collisions on both land and water more likely, despite the use of sophisticated radar and other warning devices. Ice can precipitate slips by humans and skids by vehicles. We are all aware of such dangers.

Intense heat can cause heat exhaustion and intense cold can cause hypothermia. Those suffering from either condition need help and treatment and, in the early stages, are less able to cope with other hazards which their jobs may expose them to. The hazard of being struck by lightning has to be a very rare one but there is nevertheless some risk from this natural hazard. In certain locations, where conditions conducive to the attraction of bolts of lightning are encountered, it is prudent to delay work until the storm has passed.

*15. Radiation assault (including noise and vibration)*

Into this category fall many different types of radiation. Some examples are given below.

Ultra-violet radiation from welding, plasma torches and lasers.

Infra-red radiation from hot furnaces, molten metals or glass.

Ionising radiations from X-Ray machines and radio-active sources used for a variety of purposes (thickness gauges, static eliminators etc, radiography of welds).

Microwave and radio-frequency radiation from microwave ovens, dielectric heaters, welding, diathermy, broadcasting.

More for convenience than any scientific reason, the hazards from noise (which can cause deafness) and vibration (which can cause VWF — vibration white finger) are included in this category.

*16. Electricity*

An electric current above certain quite low levels is capable of killing people. The chances of surviving an electric shock will depend on a number of factors — the state of health of the victim; the path of the electric current through the body; the moistness of the skin and so on.

Apart from the frequently, but not invariably, fatal effects of a shock, there is another hazard — that of electric burns which are deep seated and take a long time to heal.

Lastly, of course, there is the indirect hazard precipitated by electric shock where a person falls from a height and is injured.

*17. Pressure (high or low)*
Some work involves the worker being subjected to higher than normal barometric pressure. Tunnelling work and diving are perhaps the best examples. The care needed is during the decompression of those concerned to prevent the formation of bubbles caused by the release of dissolved nitrogen from the blood and the dangers to which this can give rise.

Examples of work in which ambient pressure is lower than normal to a significant extent are very few. Perhaps high flying military aircraft and work in the rarefied atmosphere on the tops of high mountains are the only ones.

*18. Not elsewhere specified*
It is always prudent in the compilation of such a list as the one given above to allow for possible omissions. It is believed that most workplace hazards can be put into one or other of the seventeen categories above, even though with some it may take a little squeezing. If squeezing does not help category eighteen remains.

It must by now have become abundantly clear that the subject of this chapter could have been the subject of a separate book. Each chapter could have been expanded extensively. This would have defeated the object of the book which has been written to prompt people to think carefully about their own workplaces. Everyone should know his or her own workplace best. This chapter is intended to stimulate discussion and debate about dangers which, because they have not actually been experienced, could so easily be overlooked.

Out of the discussions it is hoped there will emerge a better understanding of the risks inherent in the work undertaken. With a better understanding will come better planned and safer systems of work, and better, more effective controls of all the hazards identified.

## Summary of agencies causing injuries to people

1. Gravity, loss of balance (slipping, tripping, falling, striking against)
2. Falling objects and projectiles

3. Manual handling
4. Hand tools
5. Moving parts of machinery, plant and apparatus
6. Vehicles (track and free moving)
7. Fire
8. Explosions (pressure, chemical, dust)
9. Chemical assault
10. Contaminated atmosphere (including oxygen deficiency)
11. Biological assault
12. Animal assault (including humans)
13. Drowning
14. Natural phenomena (storm, wind, ice, fog, heat, cold, lightning)
15. Radiation assault (including noise and vibration)
16. Electricity
17. Pressure (high or low)
18. Not elsewhere specified

# Chapter 8
## Hazard assessment

Two components are present in any risk/hazard assessment — the probable frequency of encountering particular hazards and the possible severity of the consequences if they are encountered. Simple frequency and severity of rating tables assist in the formation of a realistic perspective and a logical way of allocating priorities.

Having assembled the list of hazardous agencies capable of causing injury to people, it is necessary to identify where each may be found in the operations being undertaken. Because resources are limited and not all things can be dealt with simultaneously (nor do they all merit the same degree of urgency and attention in any case) it is vital to have some simple method of assessing the probable frequency and potential severity of each of the identified hazards. It is always important to deal with the most serious problems first. The following method is suggested as one way of doing this.

## Frequency rating

To create a frequency rating a frequency category number should be allocated (see Figure 1). It will be noted that there are six categories. This is a quite deliberate choice. Although the distinction between one category and the one to either side of it may sometimes be difficult to determine with precision, the temptation to use a scale with an odd number of categories, such as five, has been avoided. The reason for this is that there is a noticeable tendency for us all, when we use a scale, to feel cosier in the safe middle ground. There is something comforting about the allocation of a three rating in a five point scale. With six rating options you have to make a deliberate choice towards one end of the scale or the other. You cannot prevaricate or sit on the fence.

Allocating categories could be a one-man operation but it is always better to make it a group activity. Individuals' different experiences and the debate between people of different backgrounds and disciplines, result in a more realistic conclusion. It also has the inestimable value of creating and fostering better and fuller hazard awareness by talking and arguing about it.
The following six frequency ratings are suggested:

(a) A highly improbable occurrence.
(b) A remotely possible but known occurrence.
(c) An occasional occurrence.
(d) A fairly frequent occurrence.
(e) A frequent and regular occurrence.
(f) Almost a certainty.

Highly improbable occurrences are those which, although

physically possible, are not believed to be very likely. They would include cases of flooding in areas where it was possible because of the proximity of a river or waterway, but were not known within living memory of those nearby.

Remotely possible but known occurrences, to use the example of flooding once again, would include cases where with high tides and strong onshore winds cases of flooding were known.

For examples of occasional occurrences, the entanglement of an operator's clothing on a radial arm drilling machine or other rotating item of machinery, would be appropriate.

For an example of a fairly frequent occurrence, why not take the case of the slipping ladder?

A good example of a frequent and regular occurrence is the slipping of people on a smooth and damp kitchen floor.

This leaves the category of "almost a certainty". What would suffice here for an example? Why not bumping into objects? Everyone does it fairly often through forgetfulness, poor eyesight, distraction or something being put in an unfamiliar place.

These are examples thought of in abstract. They can not compare with those decided upon by a group using their own experience about their own workplace to reach their own examples before embarking upon their category placing exercise.

The more rating categories used, the more difficult it becomes to produce categories that are sufficiently different from one another to enable selection to be made with unanimous agreement. Looked at from one point of view this might be considered a disadvantage and a sufficient flaw to undermine the value of the whole exercise. If what was being sought was a scientifically accurate measurement this criticism would be a valid one. It is not a scientifically accurate measurement which is being sought however; it is a grading of events in a sensible order by the use of general descriptions. Before pursuing the point further, consider the second parameter — that of severity.

## Severity rating

A severity rating is much easier to create with clear distinctions between the categories. The options range from negligible injuries at one end of the scale to multiple fatalities at the other. Negligible injuries include grazes or bruises, a speck of dust in an eye, or a minor sprain. Clearly with each of these conditions there could be complications which may even have a fatal outcome in some circumstances. A graze which broke the skin could admit micro-organisms from which a disease with a fatal outcome could result. This has happened with Weil's disease encountered by workers in mines, sewers and abattoirs, for example. The disease is caused by a spiral shaped bacterium leptospira icterohaemorrhagiae which occurs in areas where there is contact with water or soil contaminated by rats. The infection can enter the body via abrasions of the skin. To let the imagination run riot could make every minor injury a potential fatality. This is not the intention at all.

Minor injuries are those which require some form of first aid treatment and are those which find their way into the accident record book.

Major injuries are those of the sort which The Reporting of Injuries, Diseases and Dangerous Occurrences Regulations 1985 refer to as the injuries or conditions which have to be notified to the enforcing authority by the quickest practicable means. These are fractures of the skull, spine or pelvis; fractures of bones (with exceptions in the hand or foot which would be classified as minor injuries); amputations; loss of sight of an eye, or chemical or hot metal burns to the eyes; electric shock requiring immediate medical treatment; loss of consciousness resulting from lack of oxygen; acute illness requiring treatment, or loss of consciousness resulting in either case from absorption of any substance by inhalation, ingestion or through the skin; or acute illness requiring medical treatment where there is reason to believe that this resulted from exposure to a pathogen or infected material.

Events which cause the death of an individual, such as someone falling from a ladder that slips, or entering a vessel to clean it out and being overcome by the fumes there, are easy to identify.

Similarly, those incidents which are liable to cause the deaths of more than one operator, such as the failure of a structure; the spilling of a heavy load in an area where there are a number of others working; or the overturning of a vehicle, are all fairly easy to categorise. Cases where there could be multiple fatalities, including ones off-site, are clearly in the major hazard class involving loss of containment of toxic materials in quantity, or large scale explosions following plant failure of a catastrophic kind.

## Appendix I

## Frequency ratings summary

1. A highly improbable occurrence
2. A remotely possible but known occurrence
3. An occasional occurrence
4. A fairly frequent occurrence
5. A frequent and regular occurrence
6. Almost a certainty

## Appendix II

## Severity ratings

1. Negligible injuries
2. Minor injuries
3. Major injuries
4. Single fatality
5. Multiple fatalities
6. Multiple fatalities (including ones off-site)

## Appendix III

## Probable frequency/severity numerical table

| PROBABLE FREQUENCY | SEVERITY 6 | 5 | 4 | 3 | 2 | 1 |
|---|---|---|---|---|---|---|
| 6 | 36 | 30 | 24 | 18 | 12 | 6 |
| 5 | 30 | 25 | 20 | 15 | 10 | 5 |
| 4 | 24 | 20 | 16 | 12 | 8 | 4 |
| 3 | 18 | 15 | 12 | 9 | 6 | 3 |
| 2 | 12 | 10 | 8 | 6 | 4 | 2 |
| 1 | 6 | 5 | 4 | 3 | 2 | 1 |

# Chapter 9
## Effective monitoring

Out of every hazard assessment and subsequent management plan arises the recognition that the setting of standards and the establishment of safe systems of work and procedures are vital. Equally vital is the establishment of an effective system of monitoring to ensure that the standards which have been set are being complied with everywhere throughout the organisation concerned.

As already remarked in Chapter 8 there is no point in setting up standards and safe systems of working if you do not also introduce a formal arrangement for monitoring that the standards are being achieved and that the systems will remain safe.

## Effective monitoring

An eight point outline of a straightforward auditing technique to monitor health and safety performance is given below. It augments Step 7 given in the Management Action Plan included in Chapter 6. Its main characteristic is the setting out of the basic principles and stages for users to follow. Every individual organisation will need to adapt it to its own unique circumstances. It is intended as a guide only.

Before embarking upon the eight points we give a definition of a safety audit contained in the book "Safety Audits — A Guide for the Chemical Industry" published by the Chemical Industries Association:

> "A safety audit subjects each area of a company's activity to a systematic critical examination with the object of minimising loss. Every component of the total system is included eg: management policy, attitudes training, features of the process and of the design, layout and construction of the plant, operating procedures, emergency plans, personal protection standards, accident records etc. An audit — as in the field of accountancy — aims to disclose the strengths and weaknesses and the main areas of vulnerability or risk and is carried out by appropriately qualified personnel, including safety professionals. A formal report and action plan is subsequently prepared and monitored."

### Point 1 Declaration of intent

(a) Policy statement declaring purpose and objectives.
(b) Endorsement of the practice of auditing by senior management.
(c) Declaration that audit findings are to be implemented.

## Point 2 Selection and briefing of auditors

(a) Appointment of leader and other members.
(b) Arrangements for specialist help (eg: occupational hygienist not available in-house).
(c) Use of audit as an opportunity for training fresh auditors.
(d) Choosing the reporter and the method of recording the results of the audit.
(e) Use of cameras, videos, tape recorders, etc.

## Point 3 Preparation of an audit plan

(a) Consideration of previous audit report, its findings, and the action taken upon them.
(b) Timetable and itinerary.
(c) Notification of those in charge of the areas to be audited.
(d) Scope of audit listing areas to be omitted or given special emphasis eg: training; outside contractors' activities and the methods of liaising with them on health and safety matters.
(e) Rearranging auditors' commitments to enable audit to go ahead without interruption.

## Point 4 Conduct of audit

(a) Pace.
(b) Policy with interviews en route and afterwards.
(c) Agreement on plus and minus comments and any elements of competitativeness which may be incorporated into the audit system.
(d) Concluding meeting and agreement on the recommendations.
(e) Procedure for seeking feedback from those concerned in areas where there have been adverse comments.

### Point 5 Audit report

(a) Identification and precise location of every subject of comment.
(b) Clarity of comments and recommendations.
(c) Ordered presentation by both area and subject.
(d) Distinguish clearly the gravity attached to any particular finding (NB: any immediate hazards discovered during the course of an audit should not await the appearance of an audit report before they are dealt with).

### Point 6 Review of audit findings.

(a) Opportunity for those in charge of areas where there have been criticisms to comment on them.
(b) Auditors' comments should always be justifiable.
(c) Discussion and agreement on what needs to be done to remedy any shortcomings found during the course of the audit.

### Point 7 Action

Management should do the following:
(a) minute agreed action directives to those concerned;
(b) sanction the release of necessary resources to enable action to be completed;
(c) set time limit for the completion;
(d) call for report of progress from all managers affected;
(e) programme appropriate checks on changes as part of next audit;
(f) ensure training/awareness is modified as necessary to implement audit findings.

### Point 8 Follow-up

(a) Ensure Point 7 actions are completed.
(b) Modify standards and procedures as necessary in the light of experience derived from audit recommendations.

(c) Counsel/train managers falling short of their set objectives.
(d) Ensure subsequent audit covers the action taken on the previous one.

- **Audits are as much about building on strengths as on remedying shortcomings.**
- **Audits are not undertaken to find fault and allocate blame.**
- **An effective audit system benefits everyone.**
- **The concept of auditing has to be sold as an idea in exactly the same way as everything else has to be sold, ie: convincingly.**

# Information sources

**Acts of Parliament** HSE Guides to them

**Approved Codes of Practice and associated HSE Guidance Notes/HSE Guidance Notes** These are published on an ad hoc basis in one of five groups as follows: CS (Chemical Safety); EH (Environmental Hygiene); GS (General Series); MS (Medical Series); PM (Plant and Machinery).

**Health and Safety Commission (HSC) and Health and Safety Executive (HSE) Annual Reports** These refer to legislation and guidance published during the period covered.

**HSE/Industry Advisory Committee Guidance Literature** All the above may be obtained from HMSO or through booksellers. Standing orders within selected subject headings can be arranged. Details from PT 3C, HMSO Books, PO Box 276, London SW8 5DT.

**HSE Library and Information Services: Publications In Series: List** (two per year) This comprehensive list of official publications is free from HSE. Enquiries should be directed to Library and Information Services HSE at: Broad Lane, Sheffield S3 7HQ (tel: 0742 78141); St Hugh's House, Stanley Precinct, Bootle, Merseyside L20 3LZ (tel: 051-951 4000; or, Baynards House, 1 Chepstow Place, London W2 4TF (tel: 01-229 3456).

**Standards significant to health and safety at work IND(G)15L** Free from above HSE sources.

**HSE is also an information provider on PRESTEL** See lead frame 575.

**HSC Newsletter** (published bi-monthly) Copy from HSE, Room 414, St Hugh's House, Stanley Precinct, Bootle, Merseyside L20 3QY (tel: 051-951 4450).

**British Standards** BSI, Linford Wood, Milton Keynes MK14 6LE. BSI Enquiry Section (tel: 0908 221166).

**Monthly magazines and bulletins**

Health and Safety at Work, Bofoers Publishing Limited, Drayton Bridge House, 3 High Street, West Drayton, Middlesex UB7 7QT.

Occupational Safety and Health, RoSPA, The Priory, Queensway, Birmingham B4 6BS.

The Safety Practitioner, IOSH, Paramount Publishing Limited, Paramount House, 17-21 Shenley Road, Borehamwood, Herts WD6 1RT.

Industrial Health and Safety Bulletin, IRS, 18-20 Highbury Place, London N5 1QP.

Health and Safety Monitor, Monitor Press, Great Waldingfield, Suffolk CO10 0TL.

Safety Management, British Safety Council, 62-64 Chancellors Road, London W6 9RS.

Health and Safety Focus, AMI Occupational Health Limited, 28 Priory Road, Edgbaston, Birmingham B5 7UG.

Area offices of the HSE throughout the country will answer queries on any health and safety subject. The addresses are in telephone directories locally but they can also be found with a number of other useful addresses in the "Publications in Series:List" referred to above.

**Other titles in this series**

Croner's Guide to Health and Safety
Safe to Breathe?
Too Loud?
Classic Accidents

Available from Croner Publications Ltd, Croner House, London Road, Kingston upon Thames, Surrey KT2 6SR
(telephone: 01-547 3333).

# Index